EXPERIMENTAL
PHARMACOGENETICS

Experimental Pharmacogenetics

PHYSIOPATHOLOGY OF HEREDITY
AND PHARMACOLOGIC RESPONSES

HANS MEIER

ROSCOE B. JACKSON MEMORIAL LABORATORY, BAR HARBOR, MAINE

1963

ACADEMIC PRESS
New York • *London*

Copyright © 1963, by Academic Press Inc.
ALL RIGHTS RESERVED.
NO PART OF THIS BOOK MAY BE REPRODUCED IN ANY FORM,
BY PHOTOSTAT, MICROFILM, OR ANY OTHER MEANS, WITHOUT
WRITTEN PERMISSION FROM THE PUBLISHERS.

ACADEMIC PRESS INC.
111 Fifth Avenue, New York 3, New York

United Kingdom Edition published by
ACADEMIC PRESS INC. (LONDON) LTD.
Berkeley Square House, London W.1

Library of Congress Catalog Card Number 63-20570

PRINTED IN THE UNITED STATES OF AMERICA

This monograph is dedicated with best love, to my children, Hans Robert and Rolf Joerg Meier.

Preface

This monograph is an outgrowth of a chapter "Potentialities for and Present Status of Pharmacologic Research in Genetically Controlled Mice" published in Volume 2 of Advances in Pharmacology. While some of this material, with modifications, is repeated, much information on the hereditary aspects of the response to drugs in mice and a variety of other animal species has been added. It reviews heritable factors in animals recognized by the use of drugs and hereditary defects altering drug responses. Hereditary disorders that by virtue of their similarity to those of man may serve as useful models for further intensive investigation are presented in great detail regarding their physiopathology and biochemistry; also a great many "new areas" for future pharmacologic research are suggested. Therefore it is hoped that this monograph, while primarily addressing itself to research workers in pharmacology and genetics, may be of interest to investigators of problems in physiology, pathology, and biochemistry as well. In fact, because of the complexity of the subject, it has been necessary to assume on the part of the reader considerable familiarity with all of these scientific disciplines. It is hoped, however, that this will not discourage those whose backgrounds in any one of these areas are minimal.

HANS MEIER

August, 1963

Contents

	PREFACE	vii
I.	INTRODUCTION	1
II.	FACTORS INFLUENCING DRUG METABOLISM	5
	A. Physiologic Factors: Influence of Age and Sex	6
	B. Genetic Factors: Species Dependence of Drug Metabolism	8
III.	EXPERIMENTAL PHARMACOGENETICS	11

A. POTENTIALITIES FOR AND PRESENT STATUS OF PHARMACO-
   LOGIC RESEARCH IN INBRED MICE 11
   1. Genetic Control of Quality 12
   2. Differences between Inbred Strains of Mice 19
   3. Pharmacologic Reactions of Inbred Strains and F_1-Hybrids ... 21
   4. Potentialities of Certain Mouse Mutant Genotypes 32
   5. Pharmacologically Little Explored Materials 52

B. HEREDITARY CHARACTERISTICS OF PHARMACOLOGIC INTEREST
   IN RATS 78
   1. Genetics of the Rat 79
   2. Strain Differences in Drug Action 83
   3. Pharmacologic Effects of Serotonin and Reserpine in Rats 84
   4. Strain Differences in Cholinesterase Activity and Acetyl-
      choline Concentration of the Brain 85
   5. Sex-Linked Differences in Drug Action 87
   6. Cortisone Resistance in Pregnant Rats 88
   7. Cholesterol Synthesis in Rat Brain and the Problem of
      Myelination 88
   8. Hypercholesterolemia in Suckling Rats 90
   9. Hereditary Obesity in the Rat Associated with Hyperli-
      pemia and Hypercholesterolemia 91
   10. Hereditary Non-Hemolytic Jaundice 93

C. PHARMACOLOGIC STUDIES IN HAMSTERS 94
   1. Amphenone in the Golden Hamster 95
   2. Malignant Renal Tumors in Estrogen-Treated Male
      Golden Hamsters 96
   3. Polymyopathy of Hereditary Type in an Inbred Strain
      of Golden Hamster 97
   4. Spontaneous Hereditary Diabetes Mellitus in the Chinese
      Hamster 97

D.	PHARMACOLOGIC RESPONSES CONTROLLED BY HEREDITY IN THE RABBIT	109
	1. Genetic Studies in the Rabbit	109
	2. Genetic and Pharmacologic Properties of Atropinesterase in Rabbits	112
	3. Other Breed Differences in Drug Responses	112
	4. Hereditary Chondrodystrophy	113
	5. Recessive Buphthalmos	114
	6. Epilepsy or Audiogenic Seizures	115
E.	HEREDITARY CHARACTERISTICS OF GUINEA PIGS	116
	1. Inability to Synthesize Vitamin C (Scurvy)	116
	2. Corticosteroids in the Urine of Normal and Scorbutic Guinea Pigs	116
F.	GENETICALLY CONTROLLED DIFFERENCES IN CATALASE ACTIVITY IN RED BLOOD CELLS OF CATTLE, DOGS, GUINEA PIGS, AND MAN	117
G.	EXCRETION OF URIC ACID AND AMINO ACIDS IN THE DOG	118
	1. High Uric Acid Excretion	119
	2. Amino-Aciduria in Canine Cystine-Stone Disease	119
H.	IDIOPATHIC FAMILIAL OSTEOPOROSIS IN DOGS AND CATS: "OSTEOGENESIS IMPERFECTA"	120
I.	AMINE CONTENT IN ADRENAL GLANDS OF FAMILIES OF CATS AND GENERAL ASPECTS OF CATECHOLAMINE STORAGE IN ANIMALS	121
J.	INHERITED TRAITS IN FARM ANIMALS AND DOGS OF POTENTIAL PHARMACOLOGIC INTEREST	124
	1. Non-Homology of Cattle and Mouse Dwarfism	126
	2. Heredity and Variation in Domestic Fowl	128
K.	SPECIES DEPENDENCE OF INDUCED LESIONS	129
L.	POTENTIAL SIGNIFICANCE OF MUTAGENIC DRUGS TO MAN AND OTHER SPECIES	134
M.	COMPARATIVE ASPECTS OF BLOOD COAGULATION	137
	1. Blood Coagulation in Lower Vertebrates and Birds	137
	2. Coagulation Studies in Mammals	140
	3. Inhibitors	143
	4. Genetically Determined Clotting Abnormalities Due to Specific Factorial Deficiencies	143
	5. Therapy of Clotting Disorders	154
	6. Purification, Properties, and Composition of Bovine Prothrombin	156

	7.	Distribution of Coagulation Proteins in Normal Mouse Plasma	157
	8.	Separation and Purification of Clotting Factors from Inbred Mice	159
	9.	Fibrinolytic Activity of Mouse Endometrial Secretions (Uterone)	160
	10.	Endocrine Factors Influencing Fibrinolytic Activity and Integrity of Vascular System	161

IV. EPILOGUE   165

V. APPENDIX: PROCUREMENT OF ANIMALS FOR RESEARCH   169

VI. REFERENCES   171

AUTHOR INDEX   195

SUBJECT INDEX   204

1. Introduction

A certain interdependence of drug action and heredity has long been recognized through the experiences of workers studying the genetics of bacteria, of human geneticists classifying man according to pharmacologic assays (e.g., taste), and of cytologists observing drugs that influence heredity. However, it was only because research in the broad aspects of pharmacology and genetics has been rapidly expanding in recent years, and by focusing greater attention on biochemical aspects, that a link has been established between the two disciplines. As a result a new science, termed "pharmacogenetics," has evolved. Pharmacogenetics is defined as the study in animal species of genetically determined variations that are revealed by the effects of drugs (Motulsky, 1958; Vogel, 1959). Since the total variation in the response to drugs is caused by both hereditary and non-hereditary forces, pharmacogenetics deals only with part of the modifying factors of drug action.

While mechanisms of heredity cannot operate in a vacuum—a gene must attain its expression within the framework of a material environment (Snyder, 1955)—it is sometimes difficult to separate environmental components from genetic ones. There are many environmental events which can operate prior to and after the time a gene begins to take its effect. These environmental influences also include the effect on the genetic background (the other genes) in which a particular gene attempts to exert its own effect. Among inherited characteristics of which environmental variations do not affect the ultimate expression of the end product are the blood groups of man and animals; however, the residual genetic background may modify their expressivity and penetrance.

Although the hereditary determination of drug reactions may not always be clear, the existence of hereditary influences is without a doubt. Observations in animals are advantageous because they may be subject to special analysis by breeding procedures. Since mutations

TABLE I
ANALOGOUS HEREDITARY DISEASES OF MAN AND ANIMALS[a]

Organ system and disease or anomaly	Man	Rabbit	Guinea pig	Rat	Mouse	*Peromyscus*	Cat	Dog	Pig	Cattle	Horse
Dermal system											
Albinism (albino series)	X	X	X	X	X	X	X	X			
Elephantiasis	X								X		
Hydroa aestivale	X										X
Hypotrichosis, atrichia	X	X		X	X	X	X	X	X	X	
Hypotrichosis/anhidrosis/anodontia	X										X
Ichthyosis	X				X					X	X
Keratosis	X	X									
Nervous system											
Ataxia	X	X			X	X					
Dystrophia muscularis	X				X						
Epilepsy	X	X			X	X					
Hydrocephalus	X	X		X	X						
Spina bifida	X	X			X						
Syringomyelia	X	X									
Sense organs											
Anophthalmos/microphthalmus	X	X	X	X	X			X	X		
Cataract	X	X		X	X			X		X	X
Coloboma	X	X		X	X						
Deafness	X		X		X		X	X			
Hydrophthalmos	X	X									
Retinitis pigmentosa	X			X	X			X			
Skeletal system											
Brachydactylia	X	X			X						
Calve-Perthes' disease	X							X			
Chondrodystrophia	X	X						X		X	
Harelip, cleft palate	X				X			X	X		
Oligodactylia	X				X						
Osteopetrosis	X	X									
Polydactylia	X		X		X		X	X	X	X	X
Syndactylia	X				X				X		
Split-hand	X						X				
Tibial aplasia	X				X						

[a] Nachtsheim (1958).

I. Introduction

TABLE I (*Continued*)

Organ system and disease or anomaly	Man	Rabbit	Guinea pig	Rat	Mouse	*Peromyscus*	Cat	Dog	Pig	Cattle	Horse
Circulatory system											
Hemolytic jaundice	X			X							
Hemophilia	X							X			
Hydrops fetalis	X	X			X						
Spherocytosis	X					X					
Pelger's anomaly	X	X									
Urogenital system											
Kidney anomalies	X			X	X						
Digestive system											
Absence of second incisors	X	X									
Metabolism and endocrine organs											
Adiposity, obesity	X				X						
Diabetes	X				X						
Dwarfism	X	X	X	X	X				X	X	X
Porphyria	X								X	X	

identical to man have occurred in animals (Table I), it has been possible to establish colonies that simulate human groups in their response to drugs. One of the alternate purposes of this monograph is to demonstrate the feasibility of this approach. Animals in which the exact mode of inheritance of an hereditary disorder is established and the time of onset of the disease predictable lend themselves uniquely for studies on both therapy and prophylaxis.

While differences in response to drugs between individuals of a given species as well as between species have been considered impediments to the assessment of drug action, they are actually of value in several ways: in the general understanding of the metabolic fate of drugs, in the knowledge of the function of enzymes, and in the elucidation of possible and probable modes of action. Alterations of pharmacologic responses may result from multigenic differences between species, races, groups, breeds, and strains, but they may also be due to single genes or pairs of genes (e.g., affecting enzyme activities).

In the latter instance, the presence of (a) specific gene(s) would explain the special susceptibility of (man or) animals to certain drug effects. Since gene frequencies vary greatly between species and populations the incidence of different drug responses also varies (polymorphism).

II. Factors Influencing Drug Metabolism

From a recent analysis it becomes evident that a great many factors may markedly influence drug metabolism (Conney and Burns, 1961). Some of these include the effects of drug pretreatments on the metabolism of other drugs and the influence of drug action by either stimulating or depressing the activity of drug-metabolism enzymes in liver microsomes. The interaction of drugs and (hepatic) microsomes (endoplasmic reticulum) has been admirably reviewed by Fouts (1962); he points out the types of studies which might be most valuable and calls for the cooperation of biochemists, cytologists and cytochemists, and pharmacologists. One should add the geneticists to this list since there appears to be some correlation between endoplasmic reticulum structure and the levels of certain enzymes in the microsomes; this correlation is especially good for the drug-metabolizing enzymes. A decrease in smooth reticulum has been associated with a decrease in drug enzyme activity; conversely, preliminary results indicate that drugs such as phenobarbital which stimulate microsomal drug-metabolizing enzymes also have a marked effect on smooth reticulum as seen in electron micrographs. The effects of CCl_4 on certain microsomal enzymes involved in lipid metabolism have also been associated with change in endoplasmic reticulum structure. Such studies could be extended to other enzyme systems involved in steroid metabolism (C-20 keto reduction, etc.) and glycogen storage (epinephrine); also to drugs affecting hepatic microsomal drug metabolism, both drug-enzyme inhibitors, e.g., diethylaminoethyl diphenylpropyl acetate (SKF 525A) and stimulators, e.g., phenobarbital. In addition, triparanol (MER 29) (which influences cholesterol synthesis) and 3'-methyldimethylaminobenzene (a hepatic carcinogen) may affect the endoplasmic reticulum. Other factors influencing the drug metabolism are the physicochemical properties of the drugs themselves, e.g., pH and solubilities that alter absorption and excretion, structural molecular changes that influence penetrance into microsomes, localization of drugs in certain tissues, displacement and strong or weak binding of drugs. None of these will be dealt with; however, certain physiologic factors (age and sex) will be considered in the following section.

A. Physiologic Factors: Influence of Age and Sex

Rats, guinea pigs, and rabbits are born without the ability to metabolize drugs (Williams, 1962); this specifically relates to the oxidation of drugs by microsomes and the formation of glucuronides and sulfates. In contrast, chick embryos already possess these synthesizing enzymes. This "arrangement" may have survival value since the chicken feeds freely from birth on while the newborn mammal drinks milk. It is of interest that the series of enzymes which metabolize aromatic amino acids appear in parallel with the times of reaching immunologic maturity (Knox, 1962).

In view of the fact that the guinea pig is born with a more advanced behavioral pattern compared with that of the rabbit and rat, it is noteworthy that the newborn guinea pig has the same concentration of brain amines as the adult, whereas the brain of the newborn rat has very low amounts of serotonin and catecholamines (Karki *et al.,* 1960). Regarding the questions as to whether the newborn lacks enzyme-forming systems or whether the enzymes are present in an inactivated form, the following observations are relevant: some rats show no sex difference in drug metabolism until they are 5 weeks old; it may be that sex hormones play a part in the production of enzymes. Also tyrosine transaminase which appears abruptly after birth in the rat and 12 hours later attains very high levels can be prevented from appearing by adrenalectomy at birth. This finding is interesting since adrenal cortical stimulation in the adult will also increase this enzyme very markedly (Sereni *et al.,* 1959). An example of an enzyme which is present in young animals in an inactive form relates to *p*-hydroxyphenylpyruvate oxidase (an enzyme in tyrosine metabolism); the active form of this enzyme is very low in the newborn rat but its inactive form is present in full amount and can be activated *in vitro* (Goswami and Knox, 1961). The presence of the inactive enzyme *in vivo* is proven because—if it is not fully active—a dose of tyrosine will result in the excretion of *p*-hydroxyphenylpyruvic acid. The fact that ethionine

A. Physiologic Factors

does not block the conversion means that the enzyme is not synthesized *de novo* (from amino acids) but may be derived from a large precursor which is enzymatically inactive *in vitro* and *in vivo* until treated in a particular way (Goswami and Knox, 1961). Another possibility, namely the presence of inhibitors of drug metabolism, was suggested by the studies of Fouts and Adamson (1959) on drug metabolism in livers of baby rabbits; these included oxidation of hexobarbital, N-alkylation of aminopyrine, deamination of amphetamine, hydroxylation of acetanilide, etc.

While the most striking age-dependent changes in enzyme concentrations occur during the early part of life, certain senescent changes have also been observed. However, investigations of the mechanisms of the regulation of enzyme concentrations as a function of age must be supported by data on changes in the total number and proportion of different cell types within a tissue, and also changes in the enzymatic activity of particulates and in the number of particulates per cell. For example, in a study on senescent Sprague-Dawley rats, the concentration of succinoxidase in livers decreased similarly as the number of cells assessed by the concentration of deoxyribonucleic acid (DNA). Yet, since the concentration of DNA only estimates the total number of cells per unit weight, this measurement cannot properly indicate whether or not actual changes in enzymes occur (Barrows and Roeder, 1962). A decrease of succinoxidase in the renal cortex resulting from a loss of mitochondria during aging may be considered due to changes in enzyme activity of a given particulate within the cell. There seems to be, however, at least one change, increased catheptic activity, that occurs with age in rats. Increased catheptic activity, since it was observed without a comparable change in acid phosphatase activity, suggests that the number of lysosomes per cell did not increase with age, but rather that there is an increase in the catheptic activity of lysosomes.

It is clear that there are many facets to the biology of aging (see Shock, 1962); changes in body composition, electrolyte imbalance, and alterations in cellular structure and cell growth are but a few parameters that particularly relate to metabolism. Alterations in metabolism also lead to deviations in pharmacologic responses; virtually none have thus far been investigated thoroughly.

B. Genetic Factors: Species Dependence of Drug Metabolism

It has been known for quite some time in *microbial systems* that upon addition to a bacterial culture of an antibacterial agent, two changes occur. The first is adaptive and ensues very rapidly through "opening" of pre-existing minor metabolic pathways; the second, following selection of mutant strains possessing neutralizing enzymes, proceeds more slowly. Thus evolutionary forces can give rise to populations with a frequency of resistant enzyme systems very different from those of the ancestral type. A similar situation exists in higher animals and man, and the study of polymorphic systems provides the best approach. Evidently drugs are useful tools with which to investigate fundamentals of biochemical genetics and in particular enzymes that are controlled by allelic genes. In fact, genetically controlled drug reactions not only are of practical significance but may be considered pertinent models for demonstrating the interaction of heredity and environment in the pathogenesis of disease (Motulsky, 1958). Another facet of drug metabolism studies relates to the use of a drug as a substrate to uncover new enzyme systems (Kaplan *et al.,* 1960).

Polymorphic systems in *man* have developed for many drugs with differing proportions of phenotypes in various groups and races. The heritable factors recognized in man by the use of drugs have been summarized by Price Evans and Clarke (1961); in addition, a new book on pharmacogenetics also deals with human hereditary defects causing altered drug responses (Karlow, 1962).

The working hypothesis of making predictions from *animal* studies to man—assuming that many of the attributes of behavior found in man may also be observed in animals—has proved to be useful in drug screening and evaluative procedures despite the handicaps and limitations implicit in animal studies (Irwin, 1962). However, it is clear that in the case of drug metabolism, aside from dose effects, there is considerable species dependence (Fishman, 1961). For example, histamine may undergo acetylation, methylation, and oxidative deamination; the relative extent of these processes varies with both the animal species and the dose. While the amino acid, ornithine, is not used in metabolic conjugation by most species, it is employed almost exclusively by birds and some reptiles. The rabbit is a species which deacetylates

B. Genetic Factors

acetylated amino compounds only with great difficulty; evidence that acetylated compounds undergo deacetylation was reported for acetanilide in the dog, for a series of N-acetylsulfonilamides in the chicken, and for 2-acetamidofluorene in the rat. Perhaps physiopathologic significance relates to the enterohepatic circulation of phenols: thus, chloranystenicol in rats and morphine in dogs are excreted via the bile into the intestine as glucosiduronic acids from which they are freed by β-glucuronidase and may then be reabsorbed; this process may explain the origin of intestinal tumors of rats given 4-aminodiphenyl and its derivatives. The activity of β-glucuronidase is correlated with the action of certain hormones, e.g., estrogen-dependent fluctuations are recorded for both human mammary gland and vaginal fluid, rat preputial gland, seminal vesicle, estrus cycle, and mouse and rat liver. Other drug β-glucuronidase relationships pertain to the administration of menthol and borneol: an increase in the enzymes of liver, kidney, and spleen, but not of ovary, uterus, and pancreas, occur in the dog and mouse, respectively. Oral administration of d-glucosaccharo-14-lactose in mice (and also rat liver and kidney) strongly inhibits β-glucuronidase; other sugar lactones are only weakly inhibitory (Akamatsu *et al.*, 1961). Chlorpromazine metabolism is unquestionably complex; the total number of chlorpromazine metabolites is estimated to be close to 24. Both qualitative and quantitative differences are noted in the urinary metabolite pattern of man and dog (Goldenberg and Fishman, 1961); humans tend to favor the excretion of polar derivatives along with one or two major non-polar metabolites while dogs excrete less polar material, and the "blue" series is completely absent from dog urine. A unique feature of mouse liver has recently been recorded relative to the enzymatic degradation of azaserine; while absent from rat and pork liver, mouse liver contains an enzyme, serine-O-esterdeacidase which may be part of a detoxification mechanism and may give rise to toxic O-esters, possibly even azaserine (Jacquez and Sherman, 1962).

Recently, analogous to bacteria, mammalian somatic cells in *tissue culture* revealed differences in drug responses. Whether or not it will be possible to correlate an abnormal response *in vitro* with a similarly abnormal pharmacologic reaction *in vivo* remains to be elucidated.

III. Experimental Pharmacogenetics

Animals for research are used increasingly all over the world. The greatest numbers, especially mice, rats, and guinea pigs, are required for routine diagnosis of disease, bioassay of therapeutic substances, and screening of possible new remedies. For these and other important purposes animals are indispensable. With certain of the small laboratory animals, i.e., mice, rats, guinea pigs, and rabbits, the nature of many of the current purposes has created special requirements; they should conform with definite specifications and often be of a particular genetic constitution. In the process of breeding, species have become more or less profoundly modified, with differences becoming apparent when subjected to experiment. Generally, the rising demand is for animals of specified genetic constitution in order to obtain uniformity of results. An extreme example relates to inbred mice (Meier, 1963).

A. Potentialities for and Present Status of Pharmacologic Research in Inbred Mice

According to a current official listing by the Committee on Standardized Genetic Nomenclature of Mice (1960), 199 inbred strains and sublines of mice exist. There are sixteen inbred strain production colonies at the Roscoe B. Jackson Memorial Laboratory: six of these (A/He, AKR, BALB/c, C57BL/6, C3H, and DBA/2) were designated by the Mammalian Genetic Committee of the Cancer Chemotherapy National Service Center as Trust Stocks essential to cancer research. Demands are sufficient for these strains to justify their maintenance and one might predict an increasing interest in the diversity of inbred strains; fluctuations in demand for any one type, of course, depend upon shifts in research emphasis. Approximately twenty additional strains are maintained in the various research colonies at the laboratory.

Almost all of the standard inbred strains were established prior to 1929. The history of the development of certain inbred strains of mice has been described (Heston, 1949) and is presented graphically (Graph I). Although any one of the large numbers of strains is of special interest, in the following their specific characteristics (morphologic, physiologic, and biochemical) will be discussed and particular attention focused on pharmacologic reactions.

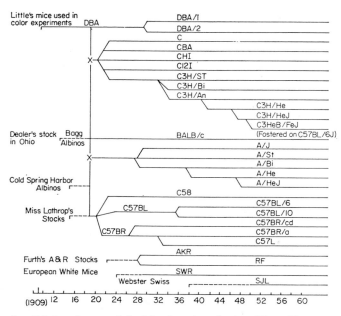

GRAPH I. Origins of some of the inbred strains of mice. (From Heston, 1949.)

1. GENETIC CONTROL OF QUALITY

No two mice are exactly alike in all respects; this phenotypic variability is due to both genetic and non-genetic variance. Although environmental variance may not be completely eliminated, it can be greatly reduced by proper design and execution of an experimental plan. Genetic variability is controlled by inbreeding; a strain is said to be inbred only after twenty or more generations of exclusive brother-sister matings. After twenty generations of inbreeding, the probability of heterozygosity is less than one percent in the absence of mutation or of selection favoring heterozygotes. An inbred strain is thus a group of mice sharing common characteristics which make them distinguishable from mice of other inbred strains. Although usually inbreeding reduces phenotypic variance, for some characteristics it may increase it; however, this variability will be almost completely non-genetic.

While theory indicates that the genetic variance within separate lines will be reduced by inbreeding, one may often want to know whether genetic causes of variability have in fact been successfully reduced. Two types of tests are available, one biological, the other statistical. The biological test requires transplanting tissues (skin, tumors) between mice whose genetic identity is in question. If the transplants are not rejected, the mice are genetically identical for all genes which govern tissue compatibilities. The statistical tests require computing a correlation coefficient between parents and offspring or between siblings. If the parent-offspring or the sibling correlation coefficient is not significantly different from zero, the strain may be regarded as genetically uniform for genes which affect the variation in the trait or traits used in computing the correlation coefficient. A significant sibling correlation coefficient does not necessarily mean that the genetic variance is significant (E. L. Green, 1962).

In some experiments it is desired to explore the relationship between two or more variables over a wide range of genotypes. A random bred population or random bred mice from double cross hybrids may have special value in such studies, provided all variables can be measured on each mouse. If it is necessary to kill a mouse to measure one of the variables, the same mouse obviously cannot be used later to measure the other variables. For instance, even though random bred mice may provide many genotypes, they are useless for studying the correlation between radiation sensitivity and natural lifespan. The mouse irradiated to estimate its sensitivity cannot be thereafter reared to its normal age of death. Similarly, random bred mice are useless in any experiment for which the effect of observing one variable persists beyond the time of observing the second variable.

Mice from a collection of inbred strains permit the experimenter to overcome the handicap of destructive tests and tests with persistent effects. Two like-sexed littermates of an inbred strain can be randomly assigned so that x is measured on one and y is measured on the other. Thereafter x and y may be analyzed as though they were measured on the same mouse. Several variables may be observed by increasing the number of littermates and allocating one to each variable. A variety of treatments whose joint effects on $x, y, z, \ldots$ are of interest may thus be studied by using inbred mice in cases where the act of measuring $x, y, z, \ldots$ destroys or changes the mice, or in cases where $x, y, z, \ldots$ cannot be measured at any one time. Also, by using inbred mice a variety of genotypes may be examined for their joint effects on several variables. In the simplest case, suppose two variables, each destructive or having persisting effects, are of interest. Take two mice each from 25 or 30 inbred strains and observe one variable on one of the pair, the other variable on the other mouse (E. L. Green and Doolittle, 1963).

There are essentially three classes of genetically controlled types which make excellent research tools (Russell, 1960); each being available in great variety.

(1) *Inbred strains** established by repeated generations of brother-sister matings and members having almost exactly the same homozygous genotype that is different from that of another strain. Only a small proportion of the genes carried by any inbred strain have been individually identified.

(2) F_1-*hybrids* obtained by crossing two different inbred strains; all members of a single F_1-hybrid type are identical but heterozygous for any genes differing between the two inbred strains. They combine all benefits that the heterozygous state confers in the way of hybrid vigor. It is claimed by some bio-assayists that in naturally outbreeding organisms, F_1-hybrids between two inbred strains tend to show a lower level of phenotypic variability than do the parental strains.

---

* According to the policy of the Roscoe B. Jackson Memorial Laboratory—in producing laboratory animals for sale—the following breeding systems were designated:

Inbred-pedigreed. These mice are individually identified by a serial number, have been propagated exclusively by brother-sister matings in each strain for at least twenty generations, and are usually provided only to research workers as mated pairs. Individual pedigrees are provided, on request, with all such animals.

Inbred. These mice are propagated exclusively by brother-sister matings, with no more than one generation from the last inbred pedigreed generation, and with no more than ten generations from a common pair of ancestors in the inbred pedigreed strain. Individual pedigrees are now provided.

Inbred-derived. These mice are first generation progeny of matings of inbred parents of the same strain. The parent mice are not necessarily brother and sister, but all trace to a recent common pair of ancestors.

F_1-Hybrids are also of value in transplantation and cancer research in that they accept tumor, skin, and ovarian transplants from mice of either parental strain. In the case of orthotopic ovarian transplants, the F_1-hybrid will furnish a superior maternal environment for the young, both pre- and postnatally.

(3) Animals carrying specified named genes or *mutant types*; certain of these are now produced for supply with the remainder of the genotype controlled as much as is feasible. Mice of these strains are the precision tools for studying the biochemical, physiological, and pathological effects of single genes. At the Jackson Laboratory alone there are some 70 stocks designed to place and maintain over 125 mutant genes on inbred backgrounds (Lane, 1960). A list of these mutant genes is presented in Table II.

Since genes probably interact in different ways it is unlikely that single gene effects in a multigenic character can be traced very far. Much of the following discussion is, therefore, restricted to effects of single gene substitutions since they may be traced most easily (Russell, 1960). However, it is seldom possible to predict the number of processes between the original gene action and the observed characteristic; they may be very close as in the case of the hemoglobin pattern, or far removed as in choreic behavior resulting from defective induction of the middle ear. The methods and materials for study of gene action have been reviewed (Russell, 1960). It should be recognized that the methods of physiological genetics, in general, favor recognition of unit genes with tissue-limited effects. Many gene-controlled reactions alter the metabolism of many kinds of cells; tests of functional capacity are, therefore, useful in determining the nature of intracellular processes affected by particular gene substitutions. Since quantitative evaluation of gene effects depends upon uniformity of the base line for comparison, it is desirable to have the action of a unit gene segregating against a genetically homogeneous (inbred) background. To maintain high congenicity the mutant heterozygote is repeatedly backcrossed to the strain of origin; if a mutation has occurred in a heterogeneous stock of animals, a new inbred strain may be produced by successive brother-sister matings with forced heterozygosis for the mutant allele or the mutant allele may be placed on an existing inbred background by repeated backcross generations.

TABLE II
SELECTED NAMED MUTANT GENES MAINTAINED AT THE JACKSON LABORATORY[a]

Linkage group	Gene symbol	Gene name	Stock designation (if any), genotype, and inbreeding	Reference
V	A^y	Yellow	C57BL/6J-A^y (N28)	GM2
VIII	an	Anemia	aa ($b\ an$) ($++$) (F8)	GM2
XV	ax	Ataxia	C57BL/6J-ax (N3)	Lyon 1955, J. Heredity **46**: 77
I	c	Albino	C57BL/6J-c (M) (N10)	GM2
I	c^{ch}	Chinchilla	$A^y a\ c^{ch} c^{ch}$ (F5)	GM2
I	c^h	Himalayan	C57BL/6J-c^h (N3)	Green 1961, J. Heredity **52**: 73
II	d	Dilute	DBA/2J-$d+$ (M) (F6); DBA/1J-$d+$ (M) (N18)	GM2
II	d^l	Dilute-lethal	aa (d^l+) ($+se$) (F9)	Coleman 1960, Arch. Biochem. Biophys. **91**: 300
XIII	Dh	Dominant hemimelia	($Dh+$) ($+ln$) (SEG A^v, a, b, c^{ch}) (NIB)	Searle 1959, Nature **184**: 1419
II	du	Ducky	DU $aa\ bb$ ($d\ se+$) ($++du$) (SEG p) (NIB)	Snell 1955, J. Heredity **46**: 27
—	dw	Dwarf	DW $dw+\ lnln$ (SEG a, b) (F16)	GM2
—	dy	Dystrophia muscularis	129/Re-$dy+$ (M) (F31)	Michelson et al. 1955, Proc. Natl. Acad. Sci. U.S. **41**: 1079
XIV	f	Flexed-tail	$aa\ ff$ (F13)	GM2
—	g	Low glucuronidase	C3H/HeJ gg (F89)	Paigen and Noell 1961, Nature **190**: 148
—	ha	Hemolytic anemia	C57BL/6J-ha (N6)	MNL **23**: 33
III	hr	Hairless	$cc\ hr+$ (F18)	GM2
III	hr^{rh}	Rhino	BALB/cHuDi-$hr^{rh}+$ (N5, F3)	GM2
X	ji	Jittery	JI $aa(ji+)(+v)$ (SEG fz, ln, s) (NIB)	GM2

A. Pharmacologic Research in Inbred Mice

TABLE II (*Continued*)

Linkage group	Gene symbol	Gene name	Stock designation (if any), genotype, and inbreeding	Reference
XIII	ln	Leaden	C57L/J $aa\ bb\ lnln$ (F84)	GM2
II	lu	Luxoid	C57BL/10JGn-lu (N29)	Forsthoefel 1959, *J. Morphol.* **104**: 89
III	lx	Luxate	C57BL/10JGn-lx (N27)	Carter 1954, *J. Genet.* **52**: 1
XI	ob	Obese	C57BL/6J-ob (N9)	Mayer 1960, *Bull. N. Y. Acad. Med.* **36**: 323
I	p	Pink-eyed dilution	C3H/HeJ-$p+$ (M) (F5)	GM2
—	rd	Retinal degeneration	C3H/HeJ $rdrd$ (F89)	Paigen and Noell 1961, *Nature* **190**: 148
III	rl	Reeler	RL $aa\ (lx+)\ (+rl)$ (NIB)	Hamburgh 1960, *Experientia* **16**: 460
V	Sd	Danforth's short tail	El $Re+Sd+Va+$ (NIB)	Grüneberg 1958, *J. Embryol. Exptl. Morphol.* **6**: 124
II	se	Short-ear	SEC/1Gn $aa\ bb\ c^{ch}c^{ch}\ se+$ (F63)	Green 1958, *J. Exptl. Zool.* **137**: 75
—	sg	Staggerer	$sg+$ (SEG a, b) (NIB)	MNL **14**: 21
I	sb-1	Shaker-1	FS $bb\ (p^{ch}\ sb\text{-}1\ fr)\ (p\ c^{ch}\ sb\text{-}1\ fr)$ (SEG A^w) (F12)	Deol 1956, *Proc. Roy. Soc.* **B145**: 206
VII	sb-2	Shaker-2	WA $bb\ (sb\text{-}2\ wa\text{-}2)\ (sb\text{-}2\ wa\text{-}2)$ (NIB)	Deol 1954, *J. Genet.* **52**: 562
—	Sl	Steel	C3H/HeRl-$Sl+$ (M) (F17)	Sarvella and Russell 1956, *J. Heredity* **47**: 123
—	Sl^d	Steel-Dickie	C57BL/6J-Sl^d (N4)	MNL **23**: 33
—	spa	Spastic	$spa+$ (SEG a, a^t, b, c) (NIB)	Chai 1961, *J. Heredity* **52**: 241-243
XVIII	tg	Tottering	$(Os+)\ (+tg)$ (SEG a, b) (F3)	MNL **18**: 40
X	v	Waltzer	V $aa\ (fz\ ln)\ (fz\ ln)\ ss\ v+$ (NIB)	Deol 1956, *Proc. Roy. Soc.* **B145**: 206

TABLE II (*Continued*)

Linkage group	Gene symbol	Gene name	Stock designation (if any), genotype, and inbreeding	Reference
III	W	Dominant spotting	C57BL/6J-W (N51)	Russell 1960, *Federation Proc.* **19**: 573
III	W^v	Viable dominant spotting	657BL/6J-W^v (N51)	Mintz and Russell 1957, *J. Exptl. Zool.* **134**: 207
XI	wa-1	Waved-1	BP $aa\ b_1b_1\ pp\ wa$-1wa-1 (SEG ln) (F10)	GM2
VII	wa-2	Waved-2	WA bb (sb-2 wa-2) (sb-2 wa-2) (NIB)	GM2
VIII	wi	Whirler	WI aa ($b\ wi$) (b+) (F7)	MNL **17**: 65
III	wl	Wabbler-lethal	WH $aa\ bb$ (br+) (+ul) (F12)	Dickie *et al.* 1952, *J. Heredity* **43**: 283

KEY TO SYMBOLS: F = Number of successive generations of brother × sister matings; N = Number of successive crosses to inbred strain; M = Mutation occurred in inbred strain; NIB = Not inbred; SEG = Segregating for other genes; GM2 = Gruneberg, H. 1952. "Genetics of the Mouse," 2nd ed. Nijhoff, The Hague; MNL = Mouse News Letter, a mimeographed bulletin issued by the Laboratory Animals Centre, Woodmansterne Rd., Carshalton, Surrey, England; (x y) (x y) = Genes x and y are linked.

a Compiled by Dr. Margaret C. Green for "Handbook on Genetically Standardized JAX Mice," 1962.

2. Differences between Inbred Strains of Mice

Information which provides a key or guide for the characterization of the strains has recently been summarized and is available for distribution from the Jackson Laboratory.* Also, a classified (subject-strain) bibliography of inbred strains has been compiled (originally only intended for use by the Jackson Laboratory personnel) and may be used by any interested investigator (Staats, 1954). A selected list on behavior studies includes pharmacologic investigations in which the end point is some activity of the mouse (Staats, 1958); mice designated "white," "Swiss," or undesignated were held to be non-inbred and omitted. Since a complete listing of all morphologic, physiologic, and biochemical *differences between inbred strains* would be too voluminous, only certain differences will be discussed.

Body size and growth (Chai, 1957b,c), life span (Chai, 1959; Roderick and Stover, 1961), various hematologic functions including leucopenia (Chai, 1957a), presence or absence of heteroagglutinins, e.g., against sheep and chicken erythrocytes (Stimpfling, 1960), are all inherited characters; serum transferrin type is genetically controlled (Shreffler, 1960); tissue concentrations of the enzyme β-glucuronidase are under control of a single gene (Paigen, 1959, 1960; Law *et al.,* 1952), the C3H being low and A high activity strains; three serum β-globulin types are controlled by a pair of alleles (Ashton and Braden, 1961); serum lysozyme, a natural antibody, is determined by but a few genes (Meier and Hoag, 1962b); other differences pertain to some physicochemical characteristics of hemoglobin (Russell and Gerald, 1958; Meier, 1961); a favorite among recent discoveries is the finding that C57BL/6 mice definitely prefer to drink 10% alcohol (higher alcohol dehydrogenase levels of livers), while DBA/2 mice are complete "teetotalers" (McClearn and Rogers, 1959); also differences in salt susceptibility (4% NaCl) have been observed: the IHB strain being resistant and the NH very susceptible (Blount and Blount, 1961); other differences relate to the spontaneous incidence and types of tumors; C57BL/6 skin is most resistant to chemical carcinogens; endocrine variation among strains is evidenced, for example, by meas-

---

* "Handbook on Genetically Standardized JAX Mice," Production Department, Jackson Memorial Laboratory, Bar Harbor, Maine.

urements of thyroid activity: C57BL/6 greater than C57BR/cd, greater than BALB/c, greater than A/J, and males greater than females (Amin et al., 1957). Strain differences are apparent also regarding distribution and contents of certain (hydrolytic) enzymes (Meier et al., 1962); great variation has been noted also for many behavioral (aggressiveness, sluggishness, etc.) patterns (Mordkoff and Fuller, 1959); distribution and nature of histocompatibility genes have been mentioned previously.

An important genetic component in the determination of serum cholesterol level in mice has been reported for five different inbred strains (Bruell et al., 1962); cholesterol levels (under identical laboratory conditions) ranged from 128 mg/100 ml in C57BL/6 to 208 mg/100 ml in C3H mice. [Gene differences in cholesterol level have been previously reported and particularly between obese (*obob*, 160–200 mg/100 ml) versus non-obese (70–140 mg/100 ml) controls (see below)]. One of the most striking findings was that the cholesterol level is significantly higher in all males than females.

Sexual dimorphisms relate to various morphologic characters (submaxillary glands, kidneys, adrenals) and also physiologic parameters, e.g., in the concentration of I^{131} (greater in males) by the submaxillary glands (Lacassagne, 1940; Llach et al., 1960) and in the lipid content of adrenal glands (females have considerably larger adrenals and greater amounts of lipid, cholesterol, and phospholipid) in STR/N (Silverstein and Yamamoto, 1961). In recent studies on the sexual dimorphism in lipid contents of the adrenal glands of CBA mice the following observations were made (M. S. Fisher, 1962, personal communication): (1) the lipid concentration is lower in males than females, (2) castration abolishes the difference, the lipid content increases in the gonadectomized male and decreases in the gonadectomized female, (3) androgens depress the level in both intact and castrate males, (4) estrogens raise the contents of both castrate and intact mice, (5) castration of males results in increase of adrenal weights and decrease in weight of the submaxillary gland, and (6) ovariectomy decreases the adrenal weight and increases the weight of the submaxillary gland. Histidine decarboxylase activity in the kidney of female mice is about twice that in males; most striking, in pregnancy the kidney enzyme level is elevated to about fifty times the non-pregnant values (Rosengren and Steinhardt, 1961). The relationship between histamine metabolism and pregnancy is as yet unsolved; the possible significance of induced synthesis of histamine in physiology and pathology has been discussed elsewhere (Schayer, 1961).

While strain differences are apparent in intact animals, differential effects occur also in "altered" mice, e.g., *gonadectomy* in certain strains of mice leads, in time, to development of tumors of the adrenal cortex which secrete estrogen and/or androgen (for references see Cranston, 1961). There is considerable evidence that these tumors develop as a result of increased secretion of gonadotropin following castration. Estrogen as well as certain other steroids and hypophysectomy will prevent their occurrence (however, cortisone, in amounts sufficient to cause adrenal atrophy, does not prevent their development).

Inasmuch as reserpine, chlorpromazine, meprobamate, nidroxyzone, perphenazine, and 2-amino-5-nitrothiazole have been found to inhibit estrus in intact mice, it is noteworthy that these drugs do not alter the frequency or the development of substrus (cornified cells present in vaginal smears) in ovariectomized C3H mice (C3H mice tolerate less drug than ZBC or ZA mice). Pharmacologically, this finding suggests a difference in the mechanism for control of secretion of estrogen from adrenals of ovariectomized mice, or, possibly a difference in the type of estrogen.

Although many more characters could be mentioned, certain of those listed and others will be discussed in more detail. However, *similarities among strains* should not be neglected. In comparative studies of normal and abnormal blood coagulation in inbred strains, physiologic variation in factors (e.g., "antithromboplastins," clotting enzyme inhibitors) was found to be lacking (Meier *et al.*, 1961); yet in investigations on the distribution (by continuous flow curtain electrophoretic separation and purification) of various coagulant activities strain-specific mobilities were observed (Allen *et al.*, 1962a).

3. Pharmacologic Reactions of Inbred Strains and F_1-Hybrids

In the following examples, strain-specific pharmacologic responses are briefly discussed. A short account was made recently by Meier and Hoag (1962a).

a. Strain Differences in the Response to Various Drugs

Perhaps the greatest problems posed by strain differences relate to the screening of carcinogenic substances; since an extensive discussion has been given by Boyland (1958) and reference to the differential induction of skin cancer has already been made (Section II,B,2), no further analysis will be given here.

Variabilities of the response to a great many drugs are well known. For example, Jay (1955) in determining the mean sleeping time in 12

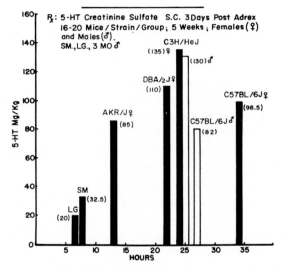

GRAPH II. Serotonin toxicity (LD_{50}). For explanation see text. (From Meier, 1963.)

inbred strains of *hexobarbital* obtained a scale of values from less than 18 minutes in SWR/HeN to over 48 minutes for A/LN following 125 mg/kg. Ambrus *et al.* (1955) found that Swiss ICR mice are 6.6 times more sensitive to the effects of *histamine* than C3H/J. Strain and sex differences to *serotonin* are given below and are illustrated graphically (Graph II; Meier, 1963). After *iproniazid* doses, high mortality and hepatic injury occurred in AKR mice, while at the same dosage there was no evidence of toxicity in C57BL and DBA/2 (Rosen, 1959). Exposure of mice of several strains to minute amounts of *chloroform* in the air results in kidney lesions that are fatal to all males but not to females; while death may follow exposure by as little as one

A. Pharmacologic Research in Inbred Mice

hour, there may be a delay of several weeks although the lesion is of the same type as in early death. Greatest susceptibility is exhibited by the following strains: C3H, C3Hf, A, HR, and DBA; the existence of a genetic difference between strains is also indicated by the finding that male mice of C57BL, C57BR/cd, C57L, and ST are resistant to amounts of chloroform that are lethal to the mice of the strains listed above (Deringer *et al.,* 1953; Shubik and Ritchie, 1953).

Reference to other compounds and *hormones* will be made in separate sections where information is provided in greater detail.

b. Strain Difference in Response to d-Amphetamine

Work that most appropriately re-emphasizes the need and importance of specifying in publications the strain of mice used in laboratory studies on drug effects has been published by Weaver and Kerley (1962). The results were obtained in investigations on the response of several strains to amphetamine and other agents; in addition (and confirmation of observations reported by other investigators), Weaver and Kerley presented evidence for greater susceptibility (lethality) of aggregated than isolated mice to the excitatory effects of amphetamine.

Mice tested were the Swiss-Webster, C57BL/6, DBA/2 strains, and BDF_1-hybrids. While differences in response to the lethal effects of amphetamine (LD_{50}, intraperitoneal, mg/kg) existed between all tested mice, and of the isolated versus aggregated (5- to 10-fold) mice, within certain strains (e.g., Swiss-Webster), no evidence for increased lethality was found for aggregated versus isolated C57BL/6 mice. Conversely, the ability of piperacetazine [2-acetyl-10-(3,4-β-hydroxylethylpiperidino)-phenothiazine], phenobarbital, and metaglycodol (2-*m*-chlorophenyl-3-methyl-2, 3-butanediol) to antagonize (reduce) amphetamine-induced lethality was evident only in aggregated Swiss-Webster mice and failed in BDF_1. However, BDF_1 mice appeared more susceptible to metaglycodol, piperacetazine, and the convulsant effects of pentylenetetrazol, but were less susceptible to strychnine than were Swiss-Webster mice.

c. Sensitivity to Chlorpromazine

Promising results have recently been obtained for the establishment of genetic bases for differences in behavioral and physiologic sensitivity to the psycho-active phenothiazine derivative, chlorpromazine. Subjects were adult mice of both sexes from C57BL/6J, DBA/2J, A/HeJ, and C3HeB/FeJ and all possible hybrids. Administration of various intraperitoneal doses (none to 4 mg/kg) one hour before a 200-second test in a photoelectric activity apparatus showed differential drug sensitivities of genotypes superimposed on a general depressant effect: with 4 mg/kg, 90% of strain C57BL/6J mice, but only 7% of strain C3HeB/FeJ failed to respond in the activity test; mice of A/HeJ and DBA/2J showed complete depression in 62 and 67% of the individuals, respectively. Results from the hybrids indicate that a simple genetic mechanism, possibly involving no more than 2 loci, could be responsible. Weight differences were not correlated with suppression of differential activity, and preliminary investigations showed similar degrees of ataxia in genotypes displaying different activity responses. The mode of inheritance and the biochemical actions involved are now being examined (Huff, 1962).

d. *Response of Inbred and F_1-Hybrid Mice to Hormones*

Clear evidence for a genetic basis of hormone response in mice has been presented by Chai (1960) and others. Although the work of Chai concerned mainly the sensitivity of response to hormonal substances, the conclusions, based on the principles of bioassay, may be applicable to other drugs and biological preparations (and to animal species other than the mouse). The following characteristics of the response were considered in the design of the assays and their computation: (1) (a linear) relationship between dose and response over the widest possible dose range, (2) response independence of standard deviation, and (3) minimal value for the ratio S^2/b^2 in quantitative assay and $1/b^2$ in quantal assay, where S^2 is the within-group variance and b is the slope of dose-response line. The latter (b) is a measurement of sensitivity; magnitude of response at a given dose may also be indicative of sensitivity, but only when comparing dose-response curves which originate from approximately the same point and are linear after appropriate transformation. In responses of C57BL/6J, C57BR/cdJ, and their F_1-hybrid, BBF_1, to androgen (testosterone propionate; weights of seminal vescicles in 4-week-old castrated males), there appeared to be no difference in the slope of the dose-response curves of the three.

Results of chorionic gonadotropin assays in females, one month of age, of DBA/1J, DBA/2J, BALB/c, and A/J, were as follows: the regression of the dose response was largest for the BALB/c and least in A/J; that of the F_1-hybrid (BALB/c × A/J) fell between those of the parental strains, but close to the A. The difference in slope between the two DBA sublines, although not large, seemed to show their genetic differences in the response as a consequence of separation in breeding for more than twenty generations.

Additional examples of variations (within limits set by the genotype) are the heritability of iodine metabolism in the thyroids of mice and the differences in response to thyroxine between inbred strains of mice and their F_1-hybrids as determined by iodine turnover rates (Chai et al., 1957; Chai, 1958).

Although it is not intended to enter into the controversy of the choice between inbred strains of mice or F_1-hybrids for bioassays, Chai's data show that in the majority of cases the responses of the F_1-hybrids to various hormone stimulations fell between those of their parental strains; this would indicate that the F_1-hybrids are not, in general, preferable to inbred strains in hormone assays insofar as the b value is concerned, for b is one of the two parameters in quantitative response and the only one in qualitative response in determining the precision of the assay. He noticed that the response to goitrogen and thyroxine in the C57BL/6J mice was distributed linearly throughout all the dosages applied, whereas the linear portion for the F_1-hybrids BAF_1 (C57BL/6 × DBA/2) fell within a limited dose. Others have emphasized that F_1-hybrids show a lower phenotypic variability than do the parental strains (Biggers et al., 1960), e.g., greater uniformity in growth and development. These are certainly desirable features in bioassay. Obviously, depending on the genetic background, some inbreds or F_1-hybrids may be more suitable for certain assays than for others, and animals of some genotypes may be less variable in one environment than in another; unfortunately the choice of assay animals cannot always be made in advance and is based either on existing data or predetermined experimentally. However, it is clear that unlike the comparison with inbred strains or F_1-hybrids, no general law can be asserted regarding random bred stocks.

In the course of the development of an assay for adrenal cortical hormones, it was noted that adrenalectomized mice (maintained on salt solution) responded to adrenaline injection in two different ways; some showed a decrease in eosinophiles and others an increase during a period of 3 to 4 hours following subcutaneous injection (Speirs and Meyer, 1949). On the basis of the ability to live after removal of the extra salt, it was suggested that adrenaline could be used to determine whether a mouse had functioning adrenal cortical tissue; a very definite correlation was found between the eosinophile responses and the presence of adrenal accessories. In addition, since the eosinopenia response to a second injection of adrenaline becomes refractory (similar to rats, dogs, and man), adrenaline pretreatment precludes a response to adrenocorticotropic hormone (ACTH), histamine, etc., but not to 11-oxycorticosteroids (cortisone) with which a quantitative response is obtained (Speirs, 1953). It is well known that accessory adrenal cortical tissue is present in many mammals. Although knowledge on presence or absence, incidence, etc., of accessory nodules should be required by those using animals for experiments involving adrenalectomy, pertinent information is available only for inbred mice and certain of their hybrids. A study of nine different strains revealed that the incidence varies (most unilateral and left side) with the strain and is higher in females than in males (Hummel, 1958). Accessory nodules were found in 50% of the inbred mice examined and in a much lower proportion of the hybrids. In mice of C57, C58, and BALB/c, there were incidences higher than 50%; in mice of strains C57BL/6, DBA/2, RIII, and CE, incidences are approximately 50%; and in mice of strains A and C3H, incidences are less than 50%. In all strains, incidences were proportionately higher in female than male adults, and in C58, BALB/c, and A incidences were higher in weanling mice. Strain, sex, and age differences could not be correlated with other known differences in these strains of mice. Thus, males of strain C3H would be the animals of choice in experiments where total extirpation of cortical tissue is required; for experiments on the functional capacity of accessory cortical tissue, females of strains C57L, C58, or BALB/c might be selected. Recently it has been found that adrenalectomized rats and mice are several times more sensitive to reserpine and serotonin (Garattini *et al.*, 1961b and personal communication); most rats are genetically heterogeneous and no knowledge exists regarding presence or absence and incidence of accessory adrenal glands. Differential susceptibilities of different strains and sexes of mice to 5-hydroxytryptamine are illustrated (Meier, 1963; Graph II).

While the evidence thus far points to differences in thyroid and adrenal function, the finding of heritable characteristics of growth rate, protein turnover, and nitrogen requirement suggests also that strains of mice differ in their ability to secrete pituitary growth hormone or in their sensitivity to this hormone. Since there is no direct chemical method for the determination of growth hormone, and cardiac glycogen, within limits, is an indicator of growth hormone activity (J. A. Russell and Bloom, 1956), cardiac glycogen levels were determined in two strains of mice. It was found that at room temperature fasting lowers the cardiac glycogen of the mouse and large doses of growth hormone are required to maintain the level observed in fed mice. At 30° C mice of the A strain increase their cardiac glycogen during a 24-hour fast, and growth hormone increases the glycogen level still further at this elevated temperature; fasting lowers the cardiac glycogen of the I strain mice, but growth hormone raises the glycogen above the level of fed animals. It was suggested that the fasting stimulus, which requires the mediation of the hypophysis and growth hormone (Adrouny and Russell, 1956), evokes a greater secretion of growth hormone in A strain mice or that the tissues of these mice are more sensitive to the hormone (Fenton and Duguid, 1962).

e. *Insulin Resistance*

High insulin tolerance of KL mice was first reported by Chase *et al.* (1948); mice of this strain can survive 300 times the doses of insulin that would "normally" be lethal. The mechanism which affords KL mice insulin resistance has been thoroughly studied (Beyer, 1955). It has been found that the carbohydrate metabolism reacts normally to the administration of insulin and that diaphragms from controls (Lt mice) did not differ in glucose consumption and glycogen synthesis with or without insulin *in vitro*. The point of difference between the KL and Lt strains is in their ability to recover from changes in carbohydrate levels upon injection of insulin; no detectable insulin can be isolated from their urines. Livers (water-soluble fraction) of KL mice contain an enzyme "insulinase" (a mixture of proteolytic enzymes?)

that displays an *in vitro* activity great enough to account for the inactivation of 200 to 300 units of insulin injected, thus enabling the compensatory reactions of adrenal hormones to become effective.

Quantal differences to insulin between strains also relate to their approximate ED_{50} convulsive responses. Among nine strains tested the C57BR/cd (650 ml unit/kg) is by far the most sensitive (also with respect to the approximate ED_{50} for histamine); the DBA/1, BALB/c, and A2G are equally sensitive (900 ml unit/kg). The response of the C3H strain differed with samples of mouse, but the difference was not investigated for significance in these determinations (950 and 1500 ml unit/kg, respectively).

f. Teratogenesis

Investigations comparing the susceptibilities of various strains of inbred mice to the same and different teratogenic stimuli are numerous. For example, in studies concerned with the production of cleft palate by cortisone in A/J and C57BL/6J, their hybrids and backcrosses showed that the genotype of both the embryo and the mother contribute to the response (Fraser and Fainstat, 1951). In the reaction to 5-fluorouracil, striking differences between strains 129 and BALB/c occurred pertaining to the periods of maximal sensitivity to certain malformations: strain 129 was more sensitive to low doses and since the period of peak sensitivities to foot malformations (poly-, macro-, oligo-, syn-, and hypodactylia) seemed to occur later in BALB/c's than in 129's, it was suggested that the former reach a corresponding stage of development somewhat later than the 129's. Pharmacologically it is of course important to investigate the mechanisms by which embryos are deformed following teratogenic treatment (Dagg, 1960).

Considering the responses to 30 and 40 mg/kg in 12-day-old embryos, the two strains were approximately the same. However, strain 129 was more sensitive at all ages to a dose of 20 mg/kg. Tentatively, it was suggested that, at low doses, effective teratogenic concentrations of fluorouracil do not reach the BALB/c embryos, or that these embryos can readily inactivate the drug. If the circulating levels were the critical factor, then the subthreshold levels in the BALB/c's may be due to a slow rate of uptake from the peritoneal fluid or to rapid rates of catabolism or excretion.

Another point of pharmacologic pertinence relates to the routes of action when the same deformity is produced by different treatments. Experimental data on a variety of treatments with relatively acute effects upon the axial skeleton in the region of the thorax have been described in detail (Runner and Dagg, 1959); interpretations of the common effects of teratogenic agents and metabolism of ectoderm and mesoderm concerned with skeletal morphogenesis have been discussed in terms of enzyme substrate associated with oxidative metabolism.

g. Differential Drug Induction of Tumors

Urethan (ethyl carbamate) has been reported to augment the induction of lymphoid leukemias by X-rays, estrogen, and cholanthrene (for review see Doell, 1962). Since urethan alone did not augment the incidence of tumors in low leukemia strains, it was suggested that it be classified as a co-leukemogen. The drug has not been found to affect the incidence of leukemia in the high leukemia strains AKR and C58; however, in Swiss albino mice injected at birth as well as in adults given toxic doses in the drinking water, urethan augmented the leukemia incidence and shortened the latent period. Recently it was shown that urethan administration to newborn C57BL/6 mice induces thymic lymphomas with a frequency approached only by that following divided doses of whole body irradiation.

Urethan is also known to induce pulmonary tumors in certain strains of mice; among them is the Bagg albino mouse. Reciprocal lung grafts from Bagg albino mice and DBA (1 day old) to their F_1-hybrids treated with urethan revealed that susceptibility to the carcinogenic action of urethan is determined by the intrinsic (genetic) properties of the lung tissue rather than by the host (Shapiro and Kirschbaum, 1951); the grafted Bagg albino lungs became tumorous. Among the factors affecting pulmonary adenoma production by urethan, data have been presented that young, more rapidly growing mice (Swiss mice) are greatly more responsive than those just arriving at maturity. It has been suggested that urethan brings about the adenomatous state by acting upon the nucleus of alveolar living cells (Rogers, 1951).

Of considerable interest is the induction of pulmonary tumors by oral isoniazid and its metabolites in BALB/c and dilute mice (*dd*). All those mice (BALB/c) receiving isoniazid or hydrazine sulfate developed pulmonary tumors (adenomas and carcinomas), but only 19% of those receiving the sodium salt of isonicotinic acid developed tumors (Biancifiori and Ribacchi, 1962). If in these mice, analogously to the rat and ox, isoniazid is converted to isonicotinic acid and hydrazine (which is degraded to ammonia), it is possible to assume that tumors observed after isoniazid doses are due mainly to the liberation of hydrazine. Hydrazine degradation may occur at different rates according to the mouse genotype.

Of dilute (dd) female mice injected subcutaneously with 2.5 mg of 4-nitroquinoline-N-oxide in 10 divided doses, starting at 2–3 months of age, all those surviving 225 days developed multiple pulmonary tumors; most of them were designated as cystadenomas but some were also metastasizing adenocarcinomas (Mori and Yasuno, 1961). The compound was previously found to be a powerful carcinogen, a single application producing skin tumors (Nakahara et al., 1957; Takayama, 1960). The production of lung tumors indicates that 4-nitroquinoline-N-oxide has also a distant carcinogenic action from the site of injection; the compound either reaches the lung directly and in sufficient amounts or a metabolite (through blood or lymphatics), which is as yet unknown, may be responsible for tumor induction (Mori, 1961). The specificity of the genotype in the tumor response has not been pointed out.

Among the various spontaneous and induced experimental tumors, considerable progress has been made in the study of their inheritance. Extensive analysis of crosses between the high tumor strain A and the low tumor strain C57BL established that multiple genetic factors were involved, the tumors appearing when the combined effects of genetic and non-genetic factors surpass a physiological threshold (Heston, 1942a,b). These studies led to the discovery of associations between specific known genes and the occurrence of pulmonary tumors. Seven genes on five different chromosomes have thus been identified; these include hairless (hr) on chromosome III, lethal yellow (A^y)* on V, shaker-2 (sh-2), vestigial tail (vt), and waved-2 (wa-2) on VII, fused (fu) on IX, and flexed tail (F) on XIV. Chemical carcinogens are

---

* The A^y also increases susceptibility to spontaneous hepatomas in males of (C3H × Y)F_1-hybrids (Heston and Vlahakis, 1961).

potent non-genetic factors which greatly increase the probability of the occurrence of the tumors above that established by genetic factors. Compounds studied included dibenz[*a, h*]anthracene, 3-methylcholanthrene, urethan, nitrogen, and sulfur mustard (Heston, 1950); radiation (Lorenz *et al.,* 1946) and recently the high concentration of inhaled oxygen (increasing the number of pulmonary tumors in strain A mice injected with dibenz[*a,h*]anthracene over that in those likewise injected but kept in air) have been studied. For a review of the differential gene actions (increasing or decreasing the pulmonary tumor incidence) see Heston (1956).

h. Aging and Spontaneous Cancer

Barring accidents, infections, etc., lifespan is genetically determined and dependent upon the strain; another determinant is the occurrence of spontaneous cancer. Differences among inbred strains relative to lifespan and tumor incidence would provide excellent ground for studies on the aging process and its relation to cancer. An illustration of an interesting approach follows: in testing the concept that endogenously produced free radicals, such as HO and HO_2, contribute to both the aging process (Harman, 1956b, 1957) and the incidence of spontaneous cancer (Harman, 1956a, 1961), various reducing substances (free radical inhibitors) were incorporated into mouse diets.

Antioxidants in two experiments had been found to prolong the normal lifespan of mice. 2-Mercaptoethylamine hydrochloride (1% w, incorporated into a pellet diet) prolonged the half-survival time of C3H female mice from 14.5 to 18.3 months, an increase of 26%, while hydroxylamine hydrochloride (1% w) produced a slight prolongation, 7%. Cysteine hydrochloride (1% w) and hydroxylamine hydrochloride (2% w) increased the half-survival time of AKR male mice from 9.6 to 11.0 and 11.2 months, respectively, a prolongation of about 15%; ascorbic acid (2.0%) and 2-mercaptoethanol (0.5%) did not have a significant effect.

None of the antioxidants studied, 2-mercaptoethylamine hydrochloride (1% w), 2,2'-diaminodiethyl disulfide (1% w), and hydroxylamine hydrochloride (1 and 2% w), prolonged the life of Swiss male mice. Hydroxylamine hydrochloride (1 and 2% w) produced a marked decrease in the tumor incidence of C3H female mice. This latter finding suggests the possibility of prophylactic cancer chemotherapy by this or other anticancer agents. Similarly, encouraging results with reducing agents against Ehrlich's ascites tumor have been reported: in addition to the aforementioned compounds N-methylformamide and potassium arsenite were demonstrated to produce tumor inhibition.

i. Neurochemical Strain Differences

Strain variations in many behavioral patterns have been mentioned previously (see above). Specifically in two strains, the C57BL/10 and BALB/c which have been carefully investigated, consistently similar results have been obtained: the C57BL/10 shows more exploratory activity, less emotionality, and superiority in fighting than the BALB/c. In a report on total, pooled brain serotonin levels, C57BL/10 averaged 1.162 μg/gm of brain tissue and the BALB/c 0.995 μg/gm (Caspari, 1960). In view of the fact that the BALB/c strain has a heavier brain and because serotonin is selectively concentrated in the "limbic" system, expression of serotonin per total brain or per gram of brain may be misleading. With this in mind, microanalytical determination of serotonin and norepinephrine levels on dissected portions of brain and consisting of diencephalon, mesencephalon, and pons were performed (Maas, 1962). The BALB/c strain had 1.34 μg of serotonin per gram of dissected brain and the C57BL/10 had 1.07 μg/gm, the difference being statistically significant ($P > 0.001$). In contrast, the values for noradrenaline (C57BL/10, 0.70 and BALB/c, 0.73 μg/gm) were not significantly different in the two strains ($P > 0.25$). While clearly the meaning of these findings is not obvious, available evidence suggests that brain amines may be related to behavioral measures. The difference noted to occur in these strains may lend itself to various experimental approaches.

4. POTENTIALITIES OF CERTAIN MOUSE MUTANT GENOTYPES

While the inheritance units determining the responses described thus far and in the mice commonly used for bioassays must be assumed to be polygenes, there are cases where the action of a single gene can have a significant effect upon certain responses. Fortunately, for an increasing number of homozygous mutant genotypes the problems of procurement have already been and are being solved. For example, in the case of the obese mouse (mutant gene symbol *ob*), homozygous (*obob*) females will not breed and males only when on a restricted diet. Matings of normal females (++, non-mutant littermates) with restricted-diet males (*obob*), and homozygous females (*obob*) with restricted-diet males have been successful. In order to provide the best genetic material the breeding colony has been made isogenic (identical genetic complement except for the mutant alleles) with C57BL/6J. Another example relates to mice with muscular dystrophy (gene symbol *dy*); mice that are *dydy* practically never leave offspring. Luckily, the $Dy \to dy$ mutation occurred in strain 129 carrying in forced heterozygosis some convenient color markers. Since ovarian transplants to healthy F_1 of this strain are successful, descendants from the transplanted tissues can be recognized (by the expression of the color gene): ovaries from *dydy* females are now routinely transplanted into normal host females and mated to *Dydy* males; since *dydy* ovaries function properly, half of the offspring are *dydy* dystrophics. Identification of certain other mutant genes, e.g., diluting genes (*ln, ru; D, d, d'*), is possible by their effect on color (*ln, ru*) or morphologic markers (*Se,* long ear, and *se,* short ear).

a. Obese Mice

Although reports of spontaneous diabetes mellitus in mice and rodents, except for hereditary diabetes in Chinese hamsters (Meier and Yerganian, 1959, 1960a, 1960b), are non-existent, hyperglycemic syndromes have been described in certain stocks of mice (Mayer *et al.,* 1951). One, a mutation (*ob*), occurred in the V stock of the Jackson Memorial Laboratory causing extreme obesity (Fig. 1) and in the homozygote, sterility (Ingalls *et al.,* 1950); some obese males could be bred when kept on a restricted diet and otherwise sterile females could be made fertile by administration of gonadotropic hormones and induction of ovulation (Runner and Gates, 1954; Lane and Dickie, 1958). The other type is a "yellow" mouse such as of the strain YBR/Wi; individuals possessing the Y gene (other than being yellow coated) become obese when fed enriched diets (Silberberg and Silberberg, 1957).

Aside from a variety of pathological changes occurring especially in obese and non-obese mice of YBR/Wi, there is a conspicuous hyperplasia and hypertrophy of the pancreatic islands common to both *obob* and YBR/Wi. Differential staining using a variety of procedures (e.g., post-coupled benzilidin reaction, Gomori's aldehyde fuchsin, and chrome alum hematoxylin-phloxin) fails to reveal deviations from the normal pattern in *obob* mice (Meier, 1960, unpublished); similarly no alterations in alpha/beta cell ratios and in cytoplasmic granulations

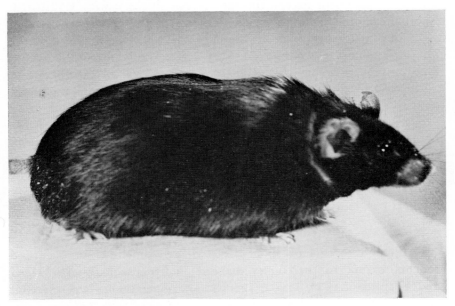

FIG. 1. Mouse with obese-hyperglycemic syndrome, *obob*. (Courtesy of Photo- and Art Department, Roscoe B. Jackson Memorial Laboratory.)

are detected in the YBR/Wi. It seems that in YBR/Wi, hypophyseal or hypothalamic dysfunction may be responsible for hyperglycemia. Both types are insulin resistant (similarly also Chinese hamsters); obese-hyperglycemic mice differ, however, from diabetic hamsters regarding hepatic glucose metabolism (see below): glucose oxidation (glucose-1-C^{14} and -6-C^{14}) to CO_2 is normal (30% of normal in diabetic hamsters); glucose incorporation into glycogen is only about ⅕ of normal (9% of normal in the hamster), while incorporation into fatty acids may be over 200% of normal (versus 15% in the hamster).

A. Pharmacologic Research in Inbred Mice

Metabolically, aside from hypercholesterolemia (see above), obese mice show an increased liver and kidney content of coenzyme A; their increased rate of acetate turnover is consistent with the elevated rate of acetylation (sulfanilamide). However, the relationship of acetylation rate to the rate of lipogenesis has still not been well defined (Thompson and Mayer, 1962). Thus factors responsible for *obesity* in O-H (obese-hyperglycemic) mice are accelerated lipogenesis and, as shown recently in studies on the effects of fasting and of adrenaline added *in vitro* to isolated epididymal adipose tissue from obese and non-obese littermates (Marshall and Engel, 1960), impaired fatty acid mobilization (lipolysis). With reference to acetate metabolism relative to lipogenesis and obesity the following observations pertain. While glucose metabolism in the isolated epididymal adipose tissue is depressed, either when expressed on a unit nitrogen basis or wet weight (see Christophe *et al.*, 1961a), incorporation of acetate carbon into fatty acids is increased in *young* obese mice (Christophe *et al.*, 1961b), suggesting increased lipogenesis as the factor responsible for obesity. However, additional studies on the acetate incorporation into fatty acids and neutral fat in the fat pads from *mature* obese mice revealed no increase of *in vitro* lipogenesis; on the contrary, lower acetate utilization was noted (Hellman *et al.*, 1962). Whether or not there is indeed a difference between developing and established obesity is difficult to decide, since the conclusions reached by Christophe *et al.* (1961b) were drawn from experiments in which no glucose was added to the medium; in the presence of glucose, lipogenesis was in fact equal in both obese and non-obese animals. Yet, the possibility of a difference between young and adult mice derives from the *in vivo* observation that young obese animals retained about three times more C^{14} in carcass lipid than did lean littermates, while mature obese mice retained only about 30% more than did mature non-obese mice (Bates *et al.*, 1955). The depressed labeling of fatty acids observed by Hellman *et al.* (1962) in mature obese mice is not likely to be due to increased dilution of acetate-C^{14}, since the pool of endogenous acetate has been reported to be of the same order of magnitude in fed obese and non-obese mice (Zomzely and Mayer, 1959).

Hellman *et al.* (1962) also comment on the fact that in both mature obese and lean mice there was a tendency for decreased acetate conversion with increasing age, an observation in accord with previous findings in the rat. The considerably higher number of fat cells per unit wet weight in lean littermates was not accompanied by any increase of tissue nitrogen; since there seemed to be a positive correlation at least for fatty acids and neutral fat and the number of fat cells, the importance of considering biochemical data with morphological findings is evident.

Since metabolism of glucose and free fatty acid (FFA) mobilization are closely interrelated, regulation of FFA release is pertinent; obesities under consideration were the hereditary O-H syndrome and that resulting from injection of goldthioglucose (Marshall *et al.,* 1955); comparison was made with their respective non-obese littermates (Leboeuf *et al.,* 1961) of FFA mobilization and glucose metabolism to glyceride-glycerol, to fatty acid, and to glycogen *in vitro*. Tissue from O-H mice metabolizes less glucose than tissue from their non-obese littermates in absence of added hormone or in the presence of insulin (0.1 unit/ml) or adrenaline ($10^{-4} M$). In addition there is also a diminished ability for insulin to inhibit and for adrenaline to augment fatty acid release (also the "fat-mobilizing factors," described recently, have no effect on adipose tissue excised from O-H mice). No such differences were observed between tissues from goldthioglucose-injected obese Swiss mice and their lean littermates. Diminished ability of tissues from O-H mice to stop release of fatty acids after *in vitro* addition of insulin may be explicable by a decreased rate of glucose metabolism to glyceride-glycerol, and the diminished effect of adrenaline on FFA release would suggest defective lipolytic mechanism. Since the release of FFA reflects a balance between glucose-dependent esterification of FFA's and the rate of triglyceride breakdown to FFA, decreased rate of FA mobilization may have major etiologic importance in this type of obesity. In order to evaluate the significance of these *in vitro* findings, growing obese mice and their littermates, both unaltered and adrenalectomized, are being chronically treated with adrenaline (one group) and chlorpromazine (Thorazine; another group); the influence of unlimited and restricted diet is being tested also. Results are not yet available (Meier, 1962, unpublished). Also, based on recent *in vitro* findings (Dole, 1961), according to which purine and pyrimidine bases as well as caffeine and pyrophosphate increased the

A. Pharmacologic Research in Inbred Mice

lipolytic action of adrenaline [ACTH, thyroid-stimulating hormone (TSH), and glucagon], groups of obese mice (both on *ad libitum* or restricted food intake) are being treated with a combination of adrenaline and caffeine benzoate hydrochloride; data will be forthcoming shortly (Meier, 1962, unpublished). Fatty acid esterification, as one of the mechanisms controlling fatty acid release from adipose tissue, will be *in vivo* the net result of a higher blood glucose (Mayer et al., 1951) and a greater stimulation by insulin (Christophe et al., 1959); and in obese animals with insulin resistance (Bleisch et al., 1952) and impaired glucose utilization, esterification will also occur. While numerous studies suggest that mobilization of FFA from adipose tissue is regulated in part by catecholamines and ACTH, only recently have data been provided to show that noradrenaline is indeed an essential factor common to the action of ACTH and possibly to other hormones (Paoletti et al., 1961); the evidence is based on (1) positive identification of catecholamines in adipose tissue [Sidman et al. (1962) have also found that noradrenaline is present in epididymal white fat and especially in interscapular brown fat of the mouse. They report values of 0.05 and 0.49 $\mu g/gm$ of wet tissue, respectively; adrenaline levels were about 10% of the noradrenaline values] and (2) after depletion of adipose tissue noradrenaline by reserpine, ATCH-induced mobilization of lipid is completely abolished. Thus it seemed pertinent to measure the relative content of catecholamine (noradrenaline) in adipose tissue of obese mice and non-obese littermates. Since daily injections of (exogenous) adrenaline and chlorpromazine had little influence on growth curves of obese mice while considerably reducing weight gains of controls, it is possible that permeability factor(s) may be operating (Marshall and Engel, 1960). The alternate approach of blocking the metabolism of (endogenous) amines by catechol-*O*-methyltransferase inhibitors (e.g., pyrogallol, guercetin) has not been tried (see Axelrod and Laroche, 1959; Axelrod and Tomchick, 1960).

In studies of metabolism of adipose tissue from O-H mice on standard and high-fat diets (saturated and unsaturated), adaptation was compared to that of non-obese mice upon substitution of one diet by the other (Lochaya et al., 1961). In tissue from non-obese mice fed the high-fat diets, glucose metabolism to CO_2 and FA's was diminished in the absence of added hormone *in vitro*, while glucose incorporation to glyceride-glycerol was increased. Under insulin stimulation (0.1 unit/ml), total glucose uptake was relatively decreased by the diets, as was glucose metabolism to CO_2, FA's, and glycogen; however, glucose carbon incorporation to glyceride-glycerol was unaltered. Under adrenaline stimulation, the sum of glucose carbon recovery was less after high-fat feeding. No effect of high-fat feeding was detected either on baseline rates of FA release or on the effects of insulin and adrenaline on this process. No differences were found between the effects of saturated or unsaturated fat diets on any parameters. The metabolism of adipose tissue from obese mice was only slightly, if at all, affected by high-fat feeding. Whether or not this lack of response is related to obesity per se, or more specifically to the hereditary O-H syndrome, will be of interest to discover.

Recently, similarities between O-H mice and mice made obese by the implantation of 11-dehydrocorticosterone pellets (compound A) were reported regarding increased lipogenesis (*in vitro*) in adipose tissue (Hollifield *et al.*, 1962). While the FFA content of adipose tissue increased with fasting in the gold-obese mice (and controls), it declined in the compound A and O-H mice. In the latter they were very low even after 2 days of fasting; depression of FFA release promotes obesity in these animals.

The factors responsible for *hyperglycemia* in *obob* mice have also not yet been completely elucidated. Degranulation of beta cells and increased pancreatic insulin content have been reported; also, despite a consistent hyperglycemia, assay of the blood-insulin level indicates hypersecretion of insulin (for references see Soloman and Mayer, 1962). Recently certain unexpected effects of alloxan on the blood-glucose level and pancreatic tissue of obese mice were recorded (Soloman and Mayer, 1962). While at 24 hours alloxan caused some swelling and vacuolization of beta cells in the lean mouse (littermate controls) and by 72 hours (following alloxan) most of the beta cells were either degranulated or contained dense granules and pyknotic nuclei, in the O-H mice alloxan caused an immediate and long-lasting decrease in blood-glucose levels and an increase in beta cell granulation (it may be added that fasting may have similar effects in normal mice). For further aspects of glucose (and amino acid) metabolism (in the liver and the diaphragm) of normal and O-H mice see Hellman *et al.*

(1961). To clarify the dispute in the literature whether or not (in steroid diabetic mammals) glutathione (GSH) is a hyper- or hypoglycemic agent, GSH (2.5 mM/kg, intraperitoneal) was found to prolong adrenaline-induced hyperglycemia in both O-H mice and their normal littermates. A very appreciable increase in plasma-glucose level occurred upon injection of syrosingopine (methyl carbethoxysyringoyl reserpate) (10–20 mg/kg, catecholamine releasing) followed 4 or 24 hours by GSH; this increase in sugar amounted in O-H mice to several hundred milligrams per cent. GSH alone had little or no effect. Chlorinsondiamine dimethochloride [4,5,6,7-tetrachloro-2-(dimethylaminoethyl)-2-methylisoindolinium chloride methochloride] (ganglion-blocking agent), with or without GSH induced in all (O-H and normal) hypoglycemia. These data are in accord with the concept that GSH-induced hyperglycemia results from potentiating adrenaline action and probably hindering adrenaline destruction; in the absence of adrenaline, GSH may have a cellular level hypoglycemic action (Beck and Liu, 1962).

Perhaps the greatest progress regarding the causes of hyperglycemia in O-H mice derives from the recent finding of an obvious hyperplasia of the adrenal cortex (Hellman and Hellerstrom, unpublished; quoted by Carstensen *et al.,* 1961); this hyperplasia (previously overlooked) is due to an enlargement of all cortical layers. The question was therefore raised as to whether or not the abnormality in carbohydrate metabolism is caused by an increased production of glucocorticoid steroids. The biosynthesis of corticosteroids has been shown to depend on the cofactor, $NADPH_2$ (nicotinamide adenine dinucleotide phosphate, reduced form) the formation of which in turn depends on adrenal phosphorylase, an enzyme that seems to be activated by adrenocorticotropic hormone (Haynes and Berthet, 1957). In this connection it is of considerable interest that an increased liver phosphorylase activity was demonstrated in the O-H mice (Shull *et al.,* 1956); while this finding is in keeping with accelerated liver glycogen turnover reported for these animals, the fact that liver hexokinase and glucose-6-phosphatase activities are normal indicate that the hyperglycemia of obese mice differs biochemically from alloxan diabetes (hexokinase and glucose-6-phosphatase are decreased). The increased liver glycogen content in O-H mice is also suggestive of excessive adrenal cortical hormone secretion. Since obesity precedes the appearance of hyperglycemia, it would seem that the hypercorticism is secondary and the result of obesity. Aside from hyperglycemia two other consequences of hypercorticism may be expected: (1) thymic atrophy (currently tested by Meier and Hoag) and (2) osteoporosis; the latter definitely occurs in mature obese mice (Sokoloff *et al.*, 1960; Sokoloff, 1960).

The biosynthesis of steroids by adrenals from normal and O-H mice was studied *in vitro* during stimulation with bovine adrenocorticotropin added to the incubation fluid (Carstensen *et al.*, 1961). The principal steroid formed was corticosterone, as evidenced by mobility in paper chromatograms, ultraviolet spectrophotometry, reduction of blue tetrazolium, and sulfuric acid chromogen formation. A small amount of aldosterone also appeared to be formed and was quantitatively determined from the blue tetrazolium reaction, since it was contaminated with an unknown Δ^4-3-ketosteroid that did not reduce blue tetrazolium. Progesterone-4-C^{14} added to the incubation fluid in one experiment was converted to both corticosterone and aldosterone. There is some evidence, although incomplete, of its conversion to small amounts of other steroids, but no cortisol or cortisone was formed. The total amount of corticosterone produced per gland was much increased in O-H mice, while the amount formed per unit weight of tissue was definitely increased in one experiment but not significantly changed in another experiment. No difference between hyperglycemic and normal mice was found with regard to the production of the reducing material in the aldosterone fraction. The fact that differences in the two experiments were observed perhaps relates partly to the use in the first experiment of older mice with a much more advanced O-H syndrome.

It is well known that insulin, which becomes rapidly bound to structural elements, can still exert its action in an insulin-free medium (Stadie, 1954). This is of interest for the work cited, since there is an increased insulin production in the O-H mice with a much higher circulating insulin-like activity (Christophe *et al.*, 1959). A binding of an increased amount of insulin *in vivo* by the adrenals of the O-H mice may result *in vitro* in increased glycogen formation in the adrenals of these mice, which can take place already during the preincubation before ACTH is added. As mentioned above, a high glycogen level may be necessary for the action of ACTH during the subsequent incubation, since it has been suggested that ACTH stimulates the corticoid formation by increased activation of adrenal phosphorylase (Haynes and Berthet, 1957; Haynes, 1958; Haynes et al., 1959).

In agreement with the reports in the literature (Constantinides et al., 1950; Reiss et al., 1953; Schoenbaum et al., 1959) that glucose takes an active part in the stimulation by ACTH of corticosteroid biosynthesis, it is possible that the combination of hyperinsulinemia and hyperglycemia in the O-H mice results in still more accentuated differences in corticosterone production between O-H mice and normal littermates *in vivo* than were found *in vitro*. The hyperglycemia in the O-H mice may be a result of increased glucocorticoid formation and may also act as a factor that perhaps accelerates the adrenal secretion of glucocorticoids. Thus there is the possibility of a vicious circle that may contribute to the occurrence of hyperglycemia in these hyperinsulinemic animals.

The appearance of radioactivity in both the corticosterone and aldosterone fractions when incubating the adrenals in a medium containing progesterone-4-C^{14} is further evidence for the active synthesis of corticosterone and aldosterone *in vitro*. In addition, this means that these steroids can be formed by conversion from progesterone. No formation of 17α-hydroxyprogesterone, cortisone, or cortisol was found. In this respect, no differences seem to exist between the behavior of the adrenals in mice and rats (Elliot and Schally, 1955; Fortier, 1959). Since some radioactivity was found in the areas of the chromatogram corresponding to 11β-hydroxy-Δ^4-androstenedione and Δ^4-androstenedione, these compounds may have been formed in small amounts. Most probably there is a deficiency of 17α-hydroxylase in the mouse adrenal, while the side-chain cleavage enzyme is capable of handling all 17α-hydroxylated steroids which are formed, thus preventing any of these leaving the cells. Investigations on the *in vitro* incorporation of uniformly labeled C^{14}-glucose in the adrenals of O-H mice (adrenal volume about twice that of normal) and non-obese littermates revealed greater glucose utilization in O-H mice; however, there was lower production of $C^{14}O_2$ and C^{14}-lactic acid per unit wet weight in O-H mice, while the formation of amino acids from glucose tended to be greater than in normal mice (Larsson et al., 1962). Glucose was utilized in the synthesis of the following amino acids: proline, alanine, aspartic acid, glutamic acid, glutamine, and arginine; characteristically large amounts of proline were formed.

It is pertinent to make reference to the effects of adrenal cortical hormones on visceral and total body fat of mice. Adult male Holtzman albino mice, on ad libitum diet, were treated with various doses of hydrocortisone, cortisone, and corticosterone for 12 days and their effect on body weight, testicular fat, visceral fat, and carcass fat was studied (Babikian, 1962). The following results were obtained: hydrocortisone at a level of 0.09 mg per animal daily had no effect on body weight and the fat depots; with 0.12 mg it resulted in a significant increase in visceral and eviscerated carcass fat with no effect on body weight as compared with the control; with 0.18 mg the visceral and eviscerated carcass fat was significantly less and the loss in body weight highly significant as compared with the control. Corticosterone at levels of 0.09, 0.12, 0.18, 0.36, and 0.72 mg, respectively, per animal daily had no significant effect on either the visceral fat or the eviscerated carcass fat. With 0.18 mg there was a significant loss in body weight, while with 0.36 and 0.72 mg the loss in body weight compared with the control groups was highly significant. Cortisone at levels of 0.09, 0.12, and 0.18 mg, respectively, per animal daily had no significant effect on fat depots. With 0.18 mg there was a highly significant loss in body weight compared with the control. It was concluded that the glucocorticoids differ both qualitatively and quantitatively in their effect on body weight and fat depots.

b. *Phenylketonuria and Audiogenic Seizures*

Mice carrying dilute genes (d, d^l), such as those of the P strain and both sublines of DBA (/1 and /2), are both phenylketonuric and subject to seizures under audiogenic stimuli. Since the gene for short ear, *se,* is closely linked to dilute (d) identification can be made at about 2 weeks of age (Fig. 2). A mechanism by which seizures may be induced is the depressant effect by alleles d and d^l of the dilute gene upon phenylalanine hydroxylase activity (Coleman, 1960). Although linkage of a factor affecting susceptibility with the dilute gene has not as yet been ruled out completely, evidence has recently been presented which indicates a positive relationship between (and direct influence of) dilute coat color and seizure susceptibility (Huff and Huff, 1962).

A. Pharmacologic Research in Inbred Mice

Previously two other possible correlations had been suggested: (1) DBA mice have the highest thyroid activity of all strains—at about one month (susceptible age) thyroid activity is twice that of non-seizing strains, dropping to about half the original level at 2 months—and

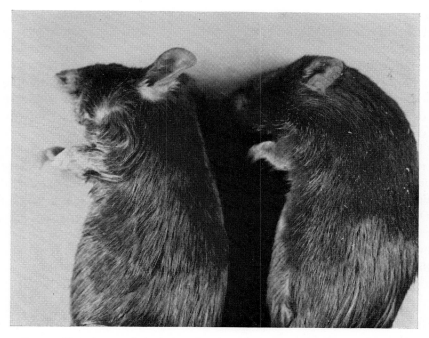

FIG. 2. Use of a morphological marker (long and short ear, respectively) to identify a particular genotype: The gene for short-ear (se) is closely linked to the gene for dilute-lethal (d^l). The mouse on the left is a double heterozygote, Dd^lSese, full-colored and long-eared; the mouse on the right is homozygous, $DDsese$, full-colored and short-eared. The first exhibits an approximately 50% inhibition of phenylalanine hydroxylase, the gene, d^l, being responsible for production of an inhibitor. Mice of the genotype d^ld^lSeSe (not shown here) are phenotypically dilute and long-eared; they are subject to phenylketonuria (85–90% inhibition of phenylalanine hydroxylase) and audiogenic seizures. The three genotypes are maintained by breeding double heterozygotes, $Dd^lSese \times Dd^lSese$.

(2) high pituitary function that induces general hormonal imbalance involving the pituitary and the adrenal cortex (and ovaries). There is proliferation of adrenal subcapsular cells with infiltration into the cortex (retention and hyalinization of corpora lutea accompanied by

hyperestrinism and endometrial and particularly myometrial hyperplasia with adenomyosis).

Unquestionably, the seizure susceptibility of dilute mice is best explained in terms of an abnormal phenylalanine metabolism; however, not all the susceptibility can be attributed to this gene (d) as other susceptible strains, A/J and HS, are DD ("intense or dense color"). Also dense backcross mice from C57BL/6J × DBA/2J crosses are often susceptible indicating that phenylalanine is only part of the over-all situation). The enzyme phenylalanine hydroxylase, which converts phenylalanine to tyrosine, was found to have only 50% of normal activity in mice homozygous for the dilute allele (d) and around 15% for the dilute-lethal allele (d^l). This decrease is caused by an inhibitor found in the particulate fraction of liver homogenates from dilute mice (DBA/1J; increases in production of phenylacetic acid paralleled decreases in tyrosine synthesis).* Thus, this situation equals a secondary or metabolic block versus a lack of enzyme in man (=primary block). Further research is now under way to test for seizure incidence among DD, Dd, and dd individuals on a common genetic background. Also experiments to determine whether or not the incidence can be altered by treatments with chemicals involved in metabolic activities related to the dilute locus are expected to clarify the role of the alleles.

Since it has been shown for man (with carcinoid tumors) and *in*

---

* Since tyrosine is the usual immediate precursor of pigments in mice and any inhibitor would be expected to cause changes in pigment formation, reduction in phenylalanine hydroxylase would readily explain their diluted pigmentation. However, studies on the absolute amount of pigment present in dilute animals indicated that there is just as much present as in non-dilute (Coleman, 1960, 1962a)—there is, however, clumping of granules. Various allelic genes at the C-locus ("albino-series"; controls intensity of hair pigmentation) decrease tyrosine (-2-C^{14}) according to C, C^{ch}, c^h, C^e, c; values in heterozygotes, Cc and Cc^h, were found to be intermediate between that of CC vs. cc or $c^h c^h$. Genic substitution of brown (bb) for black (BB) increased tyrosine incorporation twofold; A^y (agouti-locus) prevented tyrosine utilization. Recently data have been published indicating that the yellow pigment represents not simply a different degree of melanin polymerization by the tyrosinase system, but is formed, at least in part, by a different route, most likely tryptophan or a derivative (Nachmias, 1961). The genes for maltese dilution (dd) and leaden ($lnln$) affecting clumping but not the amount of pigment had no influence on tyrosine incorporation (other diluting genes, pink eyes, pp, and ruby, $ruru$, decreased tyrosine utilization). Of interest is that a newly discovered allele of the "albino-series" (Green, 1961), "Himalayan" (c^h), influences the structure of tyrosinase (Coleman, 1962a).

vitro that aromatic acid metabolites of phenylalanine are potent inhibitors of decarboxylases (Fellman, 1956; Davison and Sandler, 1958), measurements are being made in dilute mice of dihydrophenylalanine, 5-hydroxytryptophan, and glutamic acid decarboxylases (DaVanzo and Meier, 1962, unpublished). Indications that decreases of serotonin (5-HT) and 5-hydroxyindoleacetic acid in phenylketonuric patients are due to inhibition of 5-hydroxytryptophan decarboxylase by phenylalanine metabolites (Pare *et al.*, 1957) have been confirmed in normal rats fed excessive amounts of phenylalanine and tyrosine (Hess *et al.*, 1961). The amines, serotonin and γ-aminobutyric acid (GABA), have important brain functions. If phenylalanine metabolites are active inhibitors of these reactions in dilute mice, then the epileptiform seizures would not be unexpected. Seizure induction in mice with medmain (2-methyl-3-ethyl-5-dimethylaminoindole), an antiserotonin compound, has already been shown (Woolley, 1959).

Since a similar postulation, relative to decarboxylase-inhibiting effects of phenylalanine metabolites, has been made for abnormal catecholamine production [on the basis of decreased concentrations of "adrenaline-like" material in blood platelets of phenylketonuric man and inhibition of 3,4-dihydroxyphenylalanine decarboxylase (Weil-Malherbe, 1950; Fellman, 1956)], determinations of adrenaline and noradrenaline in brain, heart, and adrenals of dilute animals are also being made (by DaVanzo and Meier).* In addition measurements are being obtained of tissue dopamine, 5-HT, tryptophan-5-hydroxylase [which is the limiting reaction for 5-HT synthesis (Cooper and Melcer, 1961)], and urinary 5-hydroxyindoleacetic acid. Data have been presented of lowered noradrenaline and adrenaline in plasma, and of these and dopamine in urine of phenylketonuric children; these changes are reversed by low-phenylalanine diets. Since plasma tyrosine levels re-

---

* 5-Hydroxyindoleacetic acid is not detectable in the urine of CDBA (BALB/c × DBA/2) mice. Among tryptophan metabolites found constantly are indole, indoleaceturic acid, and kynurenine (inconstant: xanthurenic acid, kynurenic acid, and *o*-aminohippuric acid). However in CDBA mice bearing P-815 mast cell ascites tumors, 5-hydroxyindoleacetic acid and kynurenine are present in greatest amounts (also present are indole, indoleacetic acid, indoleaceturic acid, *o*-aminohippuric acid, xanthurenic acid, kynurenic acid, 3-hydroxykynurenine, and 5-hydroxytryptophan). Cytoxan (cyclophosphamide) eliminates all but indoleacetic acid, 5-hydroxytryptophan, kynurenine, indole, and 5-hydroxyindoleacetic acid (Mengel and Kelly, 1961).

mained unchanged, this decrease appears to be caused by inhibition of dopa decarboxylase; this provides further *in vivo* evidence for the relationship (or identity) of pyridoxal-phosphate-dependent decarboxylases. While tryptophan-5-hydroxylase, an intestinal enzyme, is important in the production of serotonin the question as to the significance of conversion of tryptophan to 5-HT by phenylalanine hydroxylase (liver) is doubtful (Renson *et al.*, 1961) since a mechanism for the

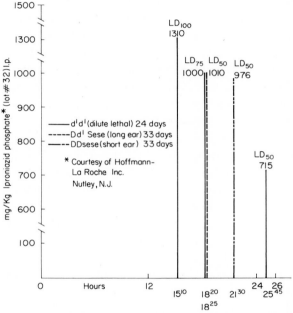

GRAPH III. Influence of genetic substitution on iproniazid toxicity in inbred mice. For explanation see text. (From Meier, 1963.)

transport of 5-HT would have to be involved (the evidence for which however is lacking). In recent experiments data were obtained that show considerable differences in iproniazid toxicity between homozygous normal animals (DD), Dd^l, and $d^l d^l$ (Meier, 1963; Graph III). Of importance is that iproniazid did not prevent convulsions, in fact they were enhanced both in time and severity [increase of catecholamines rather than serotonin (?)]; similar findings of increased sensitivity to stimuli have been reported in humans (Boshworth *et al.*, 1953, 1955, quoted by Spoerlein and Ellman, 1961). However, d,l-α-ethyl-tryptamine (Monase) precluded convulsions (Meier, 1962, personal observation).

With reference to audiogenic seizures in DBA/1 mice a protective effect of glutamic acid has been verified repeatedly (Ginsburg *et al.,* 1950; 1951; Ginsburg and Roberts, 1951); it has also been shown that the protection is substantially greater in males than females. In view of a previous hypothesis (Weil-Malherbe, 1950) relating the effects of glutamic acid to an adrenergic action, the affects of adrenalectomy (abolition) were tested (Fuller and Ginsburg, 1954). Since it was found to persist (thereby disproving the hypothesis), protection by glutamic acid is probably a central rather than a peripheral action. Interpretation of the findings may now be derived from a relationship of citric acid (Krebs)-cycle compounds to GABA (γ-aminobutyric acid); alternately, since phenylalanine is converted to phenylpyruvic acid by means of a transamination process, by increasing the concentration of available amino donor, a decrease in the phenylpyruvic-phenylalanine ratio may occur (Bowman and King, 1961).

Based on most recent observations, certain pitfalls have been uncovered in applying results of enzyme inhibition *in vitro* to functional inhibition of enzymes *in vivo*: N-(m-hydroxybenzyl)-N-methylhydrazine (NSD1034), an even more potent decarboxylase inhibitor than α-methyl-m-tyrosine [competitive decarboxylation of 5-hydroxytryptophan (5HTP) and DOPA], one hour after administration to mice (80 mg/kg, intravenously) blocks 90% of brain decarboxylase measured *in vitro*; however the rates of endogenous formation of brain 5-HT and dopamine are not affected, as measured by a rise in amines after administration of a potent monoamine oxidase (MAO) inhibitor. After very high doses of NSD1034, the decarboxylase is blocked 100% by *in vitro* test, but the rise in brain 5-HT after MAO blockade is decreased only by 50% (Hirsch *et al.,* 1962), while more than 90% inhibition of decarboxylase does not slow noradrenaline formation indicating that decarboxylation is not a rate-limiting step (Kuntzman *et al.,* 1962).

The effects in learning ability (simple mate-learning assay) of changes in the serotonin content in the brain have recently been studied by Woolley (1962). Increases in serotonin caused specifically in brain by administration of 5HTP plus the antiserotonin 1-benzyl-2,5-dimethylserotonin (BAS) resulted in complete failure to learn the maze; less specific increases caused by administration of iproniazid likewise reduced learning ability, but less than 5HTP plus BAS. Decreases in serotonin and catecholamines caused by feeding large amounts of DL-phenylalanine plus L-tyrosine increased learning ability; thus learning ability greater than normal was induced in adult mice (not specified). By contrast when a deficiency in serotonin plus catecholamines was induced in newborn mice, their learning ability when matured was found to be greatly reduced. Just as in the human disease (phenylketonuria), the time of life at which the deficiency was induced was crucial to the mental failure.

Protection against sound-induced convulsions (audiogenic seizures) was obtained in non-inbred Swiss mice by various phenothiazine ataractics (Plotnikoff and Green, 1957; Plotnikoff, 1958, 1960) while chlorpromazine was entirely inactive in an inbred strain of Swiss mice (Plotnikoff, 1960). In view of these diametrically opposed data the possibility arose that development of drug resistance in inbred Swiss may be incidental to inbreeding (Plotnikoff, 1961). Chlorpromazine in doses which had been shown to inhibit seizures in non-inbred mice and induced typical symptoms of chlorpromazine medication (ptosis, heavy sedation, and ataxia) prior to auditory stimulation exerted diminishing protection from parental to each succeeding filial generation; a similar trend pertained to analogs promazine, perphenazine, prochlorperazine, and trifluoperazine. In contrast to the phenothiazine ataractics, the barbiturate sodium phenobarbital uniformly gave protection against convulsions in all inbred generations tested. The lowered response to protective effects of chlorpromazine in each succeeding generation suggests that the genetic characteristics of drug sensitivity were lost during the course of inbreeding. It has been proposed that by utilizing differences that exist between inbred and non-inbred Swiss mice, neurochemical mechanisms may be uncovered (e.g., enzymatic intermediates that are deleted by inbreeding and essential for normal drug activity) that are responsible for drug action and drug resistance.

c. *Use of Hairless Mice*

Mice carrying the genes rhino (hr^{rh}) and (another-line) hairless (hr), an allele of rhino (Fig. 3), which produce hypotrichosis and hyperkeratosis, provide a fertile field for the investigation of factors influencing hair growth and epidermal differentiation. In a recent study the hypothesis was tested that the skin of rhino and hairless mice is unable to utilize amounts of vitamin A supplied by the normal diet, resulting in changes leading to epidermal hyperkeratinization and loss of body hair, with subsequent cyst formation of epithelial derivatives.

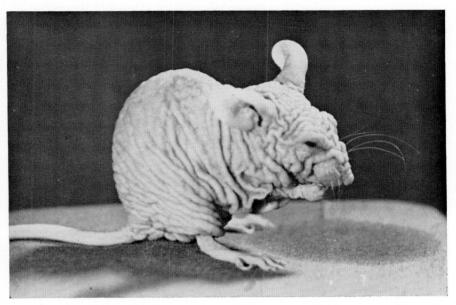

FIG. 3. Rhino mouse, $hr^{rh}hr^{rh}$. (Courtesy of Photo- and Art Department, Roscoe B. Jackson Memorial Laboratory.)

Frequently, when massive doses of vitamin A were fed to mutant mice, the beneficial effects of treatment on the hyperkeratotic process were inseparable from the pathological systemic effects induced in hypervitaminotic mice. At comparable dosage schedules, rhino mice were much more sensitive than either hairless mice or normal littermates to the systemic effects of continuous administration of vitamin A; fatty infiltration of the liver and spleen, with hepatosplenomegaly, were prominent pathological findings. These lesions may be interpreted as a more marked inability of rhinos to utilize vitamin A, or as a result of a non-specific decrease in vigor caused by the homozygous rhino condition. Other systemic effects found in treated rhinos (and normals) were hyperemia of various internal organs (lungs, kidneys, gastrointestinal tract, etc.) with hemorrhage in severely affected animals. Hairless mice were less affected systemically than rhinos, but they did exhibit similar skin responses. Both the observation that the skin of male and female hairless mice differ grossly, and the sex differences in response to treatment cannot be, as yet, directly related to the action of vitamin A. The assumption that a local lack of, or dysfunction in, the ability of skin cells to utilize vitamin A or its metabolites is responsible for the defect in hyperkeratotic mice is an attractive hypothesis. The particular metabolic block(s) involved must be rigidly specific for ectodermal cells, since, among epithelial tissues, only the skin and its appendages (including the nails) appear to be involved in the pathological process leading to hyperkeratosis. The differences between rhino and hairless patterns of hyperkeratosis (and hair loss) deserve further experimental investigation.

Actually the term "hairless" is a misnomer because these mice are born with the ability to grow a normal coat,* and during the first 2

---

* Interesting studies on the influence of alloxan diabetes, methylthiouracil, cortisone, and adrenaline on the utilization of glucose-C^{14} and L-cystine-S^{35} and mitotic activity by hair follicles in white mice are referred to briefly because of their pertinence to the general problem of hair growth, endocrine states, and drug action (Davis, 1962). Autoradiographs of hair follicles 1, 3, 4, and 8 days after the induction of growth by plucking were prepared from mice injected with 15 µc glucose-C^{14} 5.5 hours before autopsy. It was demonstrated that radioactivity increased principally in the *active* regions of the follicle, namely, the regions engaged in cell proliferation or keratin synthesis. The level of radioactivity in these regions of the follicle increased with maturation. Alloxan diabetes and cortisone induced a conspicuous depression of glucose-C^{14} adsorption during anagen (day 1, 3, and 5 follicles) but these effects were markedly reduced in metagen (day-8 follicles). Adrenaline and methylthiouracil (MTU) had no effect on the absorption of glucose-C^{14} at any of the stages of follicle activity observed. Mitotic counts in follicles from the same animals in which glucose-C^{14} utilization was studied revealed that these endocrine states exerted similar effects on cell proliferation. Alloxan diabetes and cortisone strongly depressed mitotic activity during anagen, but had a relatively weak effect during metagen; in fact cortisone appeared slightly stimulative, while adrenaline and MTU had no apparent influence. The radioactivity present in follicles representing 2, 4, 6, and 9 day stages of development was assessed by autoradiographic means from mice injected with 1 µc L-cystine-S^{35} 24 hours before autopsy. L-Cyst(e)ine-S^{35} was absorbed in large quantities only in day-9 follicles, in which it accumulated mainly in the keratogenous zone. Absorption of L-cyst(e)ine-S^{35} by metagen (day 9) follicles was unaffected by alloxan diabetes, cortisone, adrenaline, and MTU.

A. Pharmacologic Research in Inbred Mice

weeks of life, histology of the skin shows no anomaly. The mice gradually shed their hair progressing from the nose downward over the entire body until they become completely naked by the third to the fourth week. Although occasionally a certain regrowth of hair may occur, the mice remain hairless thereafter. Beginning at about 6 weeks of age (corresponding to the human age of approximately 10 years) and becoming marked by 5 months (approximately 30 years of human age), the smooth young skin becomes wrinkled and rough (Montagna et al., 1952, 1954). Recently evidence has been obtained suggesting that hairless mice lend themselves for screening of cosmetics (Homburger et al., 1961). In the past, the cosmetic industry has primarily addressed itself to the biologist for experimental evaluation of skin preparations (for testing of safety, absence of irritants, and sensitization) in rabbits and guinea pigs. These species were chosen not necessarily because of any special property but because of the standardization of conventional methods commonly employed. Clearly the availability of pure genetic lines of other species would be more ideally suited for systematic cosmetologic research. The skin of certain sublines of hairless mice mimics in many respects the aging phenomena in women, i.e., wrinkling of skin due to dehydration and connective tissue changes, and was found to respond to substances known to be effective in women and to cause smoothing of wrinkles and rehydration. Also, it was possible in these mice to differentiate between the effects of cream bases, solvents, and active ingredients, and to study any effects on organs other than the integument which may result from the application of these materials to the skin. A strain designated HR/Bio, suitable for study up to the age of about 4 months, and one designated *hrhr*/J, suitable up to about 2 months of age, were used. Of the materials tested in the study and successfully reversing the skin lesions (or at least smoothening the skin by hydration and densification of dermal connective tissues) upon application were estrogens, estrogen and progesterone, ethisterone and pregnenolone; with testosterone, on the contrary, the skin was made to look rougher. All preparations containing estrogen (except 2-methoxy estradiol) resulted in generalized smoothening of skin; a nearly similar effect was obtained with Δ^5-pregnenolone. With ethisterone, the smoothening effect was more marked at and around the site of application while an unmedicated cream base* was effective more or less confined to the area of application. Microscopically, the beneficial effects consisted of densification and hydration of the dermis with widening of its papillae and consequently stretching and flattening of the skin folds. Interestingly, the unmedicated cream base produced a desirable effect through hydration of epidermal cells only; when the hormonal ingredients of the medicated cream formula were given in solvents rather than cream base there was complete absence of epidermal swelling.

Clearly then, the study by Homburger *et al.* (1961) not only holds considerable commercial promise, but has opened up avenues for more basic investigations in cosmeto-pharmacologic research dealing with the mechanism of action rather than simply clinical effectiveness of compounds. During the course of assays, mice may be biopsied and necropsied for chemical and histochemical studies. Such investigations, e.g., concerning the water, lipid, and protein content of skin, and also gas chromatographic studies are in progress. While estrogen greatly increases the water content of skin, pregnenolone does not (Homburger, personal communication, 1962).

5. PHARMACOLOGICALLY LITTLE EXPLORED MATERIALS

In discussing important characters (with respect to pharmacologic investigations) of genetically controlled mice, certain features of transplantable tumors should also be included. In addition, since recent medical advances have increased the relative importance of constitutional or inherited diseases in man (by decreasing the prevalence and severity of infectious diseases), analogous hereditary diseases in experimental mice may be useful for study of their etiology, prophylaxis, and therapy.

a. Usefulness of Transplanted and Induced Tumors

The Tumor Bank of the Jackson Laboratory maintains some 25 or more tumors; one, a preputial (sebaceous) gland tumor, ESR 586, that is transplantable in C57BL/6J mice is of particular interest because of its relatively high content of a variety of sterols (Kandutsch and Russell, 1959, 1960a,b). About one week following transplantation, quan-

---

* The authors (Homburger *et al.*, 1961) identify the creams used as Ultra Feminine Cream of Helena Rubinstein.

tities of at least 4 chromatographically separable reducing steroids or α-ketols are excreted in the urine (Meier, 1960, unpublished); in recipient males the kidneys, adrenals, and submaxillary glands undergo a certain degree of feminization. Some 10 different sterols have now been identified. First, the occurrence of lanosterol; 24, 25-dihydrolanosterol; 4α-methyl-Δ^8-cholesterol; Δ^7-cholesterol; 7-dihydrocholesterol; and cholesterol; second, the kinetic evidence from acetate-C^{14} incorporation; and third, the studies on the metabolism of the labeled sterols in cell-free homogenates have provided evidence for the operation of a pathway that includes these sterols in the sequence given. There seems to be, therefore, an alternate pathway from lanosterol to cholesterol than lanosterol, C_{29}-sterol, C_{28}-sterol, zymosterol, desmosterol, and cholesterol. Obviously ESR 586-bearing mice would be useful for investigating inhibitory compounds of cholesterol synthesis.

In a recent study utilizing liver extracts of normal LAF_1, BAF_1, and B10.D2 (a C57BL/10 subline) mice, several hitherto unknown enzymatically catalyzed reactions have been found to exist along which cholesterol precursors can be diverted. Among these is the conversion of cholesta-4,7-dien-3-one to cholestenone which occurs in two steps: the first, catalyzed by Δ^7-isomerase results in isomerization of the Δ^7 bond to the 6 position and yielding cholesta-4,6-dien-3-one; the second step results in reduction of the Δ^6 bond to give cholestenone (Kandutsch, 1962a). It was postulated that cholesta-4,7-diene-3-one may arise as a result of the oxidation of the 3-hydroxyl of 7-dehydrocholesterol which is a known precursor of cholesterol. The reactions indicated are obviously of interest in relation to the mechanisms by which cholesterol biogenesis is regulated. The existence of enzymes capable of catalyzing some of the steps along this pathway and the presence of cholesta-4,6-dien-3-one in arteriosclerotic aortas and hog spleens suggest indeed its operation under physiologic conditions. The product of these reactions, cholestenone, is known to inhibit (upon injection) the biosynthesis of cholesterol. Also, several other compounds which alter cholesterol biosynthesis are potent inhibitors of Δ^6-reductase; these include bile acids, detergents, and MER 29. Metabolism of a cholesterol precursor to cholestenone through the two-step reactions described may thus provide a pathway for steroid hormone formation that does not include cholesterol (indeed, conversion of cholestenone to steroid hormones by hamster adrenals has been demonstrated). In view of the fact that these unsaturated steroids under study are especially susceptible to oxidation by free-radical mechanisms and, since generation of free radicals as a result of cancer chemotherapy is known to occur, these reactions relate particularly to carcinogenesis; oxidation of steroids might result in the production of (a) compounds with carcinogenic activity or might seriously alter the course of subsequent metabolic reactions (Kandutsch, 1962b).

Following early gonadectomy, adrenal cortical tumors arise at 6 months of age in mice of the F_1 generation resulting from the cross DBA/2 WyDi female × CE/WyDi and the reciprocal cross; by morphologic criteria, these tumors appear to have androgenic effects upon the secondary sexual structures (submaxillary glands, kidneys). Biochemically, however, no appreciable amounts of stored androgen or estrogen were found; also, it was found (1) that there was no conversion of progesterone-4-C^{14} to androgens or estrogens *in vitro,* (2) that this precursor was rapidly converted to corticosterone and deoxycorticosterone, and (3) that this conversion was (on a weight by weight basis) less efficiently carried out by neoplastic rather than by normal adrenal tissue (for references see Hofman and Christie, 1961). Recently it has been found that C-11 and C-12 hydroxylase activities were reduced in tumor tissue as compared to normal tissue; to a lesser extent, C-17 and C-20 desmolase activities were also reduced and 17β-ol-dehydrogenase, present in normal adrenals, was absent in tumors. Despite the fact that the neoplastic tissue did not convert the substrate 17α-hydroxyprogesterone-4-C^{14} to known steroidal androgens in any abnormal degree, much androgenic substance could be potentially synthesized *in vivo* (despite reduced enzymatic efficiency) as a consequence of the very great increase in tissue mass (Hofman and Christie, 1961).

Useful tools for studies on the mechanisms, within the pituitary, involved in the control of thyrotropin secretion are the transplantable pituitary tumors developed by Furth (Furth *et al.,* 1953). Two types of tumors, thyrotropic and adrenotropic, were used to examine this problem each reflecting quite closely the behavior of its prototype cell in the normal pituitary; these transplanted tumors are indigenous to LAF_1 mice. As control tissues for hormone assays served a spontaneous fibrosarcoma in LAF_1 mice; kidney, muscle, and pituitary from thyroidectomized LAF_1 mice, and a heart carcinoma of Swiss mice. The results suggested that there may be a special thyrotropin-secreting cell with an unusual pattern of metabolism for thyroxine. Thus, mouse pituitary thyrotropic tumors were found to contain large amounts of 3, 5, 3-triiodothyronine 4 hours after injection of labeled thyroxine, whereas pituitary adrenotropic tumors and other tissues contained none.

Two of the three thyrotropic tumor strains grow only in thyroidectomized mice; the adrenotropic tumors were maintained in both intact and adrenalectomized mice given deoxycorticosterone (subcutaneously once per week). Since a specific alteration in hormonal environment was required to induce each of the secretory types, the resultant tumors are histologically homogeneous and specific. With dependent tumor strains, the same specific end organ hormone is required to suppress growth and hormonal secretion as is normally required to suppress pituitary secretion of the corresponding tropic hormone. These close similarities to normal pituitary function suggest that each of the secretory pituitary tumors may indeed have originated from corresponding specific pituitary cells and thus reflect their functional behavior. From the results obtained with the pituitary tumors, there may not only be a specific thyrotropin-secreting cell but, in addition, it may have a unique metabolic pathway for thyroxine. The formation of large amounts of 3, 5, 3-triiodothyronine could be related to the inhibitory action which thyroxine exerts on thyrotropin secretion; since none accumulated in the serum or various control tissues, predilection of the thyrotropic tumor for 3, 5, 3′-triiodothyronine seems eliminated (Werner et al., 1961).

b. *Neuromuscular Mutants*

Although inherited neuromuscular defects in mice are relatively frequent, few have been studied thoroughly. Exceptions are the dilute-lethal (d^l) and the dystrophia muscularis ($dydy$) which is a primary myopathy without central or peripheral neural defects; both these mutant genotypes are under extensive investigation. Careful histologic examination revealed that although myelination proceeds in the same sequence in dilute-lethal mice as in normal homozygotes and heterozygotes of the d^l/Sl line, myelin degeneration (Marchi technique) was observed in the vestibulo-spinal, spino-cerebellar, and tecto-spinal systems. These structures differentiate during the 11th and 15th day of embryonic development; those that remain normal differentiate subsequent to the 15th day. Although these changes appear to be primary, it is questionable whether the altered behavior may be explained on the basis of demyelination, since some structures showing degeneration did not produce the expected functional alterations (Kelton, 1961). Among some fifty or more separate mutations affecting behavior abnormalities most concern organogenesis, e.g., failure of proper induction of the labyrinth and postnatal degeneration of the organ of Corti (Grüneberg, 1952a); these lead essentially to "circling" movements. Histopathologic alterations of the central nervous system have been observed in a few mutants. Total or partial absence of the corpus callosum (*acac*) was discovered accidentally in the course of studies of the brain anatomy of mice with rodless retina; clinically they cannot be distinguished from normal mice (King and Keeler, 1932; King, 1936). Demyelination occurs in some "staggering" mutants;* degeneration and loss of Purkinje cells, and, to a lesser degree, of mossy fibers of the cerebellum are observed in the "agitans," *agag* (Martinez and Sirlin, 1955). "Polycystic alterations" occur in the white matter of the jittery (*jiji*) mouse (Harman, 1950) and degeneration of retinal rods is encountered in several inbred strains (Tansley, 1954). Many mice that show epileptiform seizures and a miscellaneous group with various behavioral abnormalities have no lesions or they have not yet been discovered.

Obviously certain neuromuscular mutations involve biochemical rather than morphological changes; an example (dilution) has already been given. Recently ("all-or-none") enzymatic differences were found in zymograms of brains from a number of neuromuscular mutants as compared to appropriate controls (Meier *et al.*, 1962a), and certain

---

* It may be thought of interest to study free nucleotides of the brain of certain of these mutants, especially those with demyelinating syndromes, in view of the fact that they play an essential role in energy metabolism (adenine nucleotides), biosynthesis of lipids (cytosine nucleotides), polysaccharides (nucleotides of uracil and guanine), and proteins (guanosine triphosphate). Attempts at measuring free nucleotides of the cerebral hemispheres of four species of mammals including mice have been made (Mandel and Harth, 1961); separation and quantitative estimation were effected by ion-exchange chromatography. For example, ATP represents, in the mouse (normal 18–25 gm) and rat, respectively, 58 and 51% of the nucleotides; the ratio ATP/ADP was found to be 22 for the mouse and 9.4 for the rat.

primary effects of three neurological mutations on the electrophoretic patterns of serum proteins have also been reported (Yoon, 1961).

Using the zymogram technique, which combines zone electrophoresis in gels, and histochemical procedures attempts have been made to study genotypic differences among inbred strains and F_1-hybrids. Certain of the results obtained were as follows: patterns of serum β-galactosidase were similar (single band) in adult AKR/J, DBA/2J, B6D2F$_1$ (C57BL/6J × DBA/2J) and C57BL/6J, but the reaction rates in the last two (related) were much slower. The relationship of the F_1-hybrids to one of the parents also showed in the alkaline phosphatases; two bands appeared in the C57BL/6J and B6D2F$_1$ with sodium α-naphthyl acid phosphate while only one band appeared in DBA/2J. The reverse occurred using sodium β-naphthyl acid phosphate. AKR/J revealed only one band with both substrates.

To identify the sensitive cholinesterase, eserine sulfate at a final concentration of $10^{-4} M$ was applied prior to substrate addition; the possible influence on esterase activity of other drugs was also studied. Indoxyl butyrate was used as the sole substrate; tissue supernatants and serum were derived from C57BL/6 retired breeders. Carbamate ($10^{-4} M$) inhibited 2 zones in lung supernatants, and pentobarbital (60 mg/454 ml of the incubation mixture) blocked one band in each brain and kidney. The inhibition of brain cholinesterases by chlorpromazine ($5 \times 10^{-4} M$) which was also observed, has recently been confirmed for man (Johannesson and Lausen, 1961). As has been shown previously esterase activity increases as development proceeds, but in a few instances maturing tissues showed a transient decline in activity of certain esterases, e.g., losses of bands occurred in liver, kidney, and brain of newborn (24-hour old) C57BL/6J versus 19-day-old fetuses when indoxyl acetate was used as substrate (Meier, 1961, unpublished).

The inheritance of serum esterases (α-naphthyl acetate) having different electrophoretic patterns (in starch gels) was investigated by Popp and Popp (1962). They found two classes of inbred mice, one with a single band, as in C57BL and C57L, and another revealing two bands, as in 19 other strains, e.g., BA1B/c and AKR. Genetic analyses indicated that such differences are controlled by allelomorphs of a single locus, designated *Es*.

In a neuromuscular mutant, ducky [showing waddling gait and lack of coordination (Snell, 1955)], differences were discernible between ducky ($dudu$) and normal ($du+$ or $++$) in esterase activities of brain, liver, and kidneys using both indoxyl acetate and indoxyl butyrate: three bands in normal kidney versus one band in ducky (3 and 1 are identical in regard to position); two bands appeared in brains of normal versus none in ducky and five bands were distinguished in ducky liver but only four in normal (the extreme cathodal one is lacking). In a recently discovered mutant causing spastic symptoms in a random bred stock of mice (Chai, 1961), the cerebellum of the spastic homozygote ($spaspa$) contained one less esterase band (2) using β-naphthyl acetate than did the normal homozygote or heterozygote. A mutant, dilute-lethal ($d^l d^l$; see above), which shows chronic convulsions and opisthotonus until death at about 3 weeks of age (Searle, 1952), lacked one of two bands present in brains of normal mice treated with indoxyl acetate. The brain of a recessive mutant tentatively called shimmy (Lane, 1961) entirely lacked esterase activity with both indoxyl butyrate and indoxyl acetate whereas two bands are present in the normal mice. The "reeler" syndrome (Figs. 4 and 5) is caused by a single recessive mutation (Falconer, 1951). Mice homozygous for this gene (rl) lack muscular coordination, suffer balancing difficulties and tremors. Although the cerebellar cholinesterase in homozygous animals was reported to be doubled or tripled over that of normal littermates (Hamburgh, 1958), little activity was found in zymograms and only with naphthol acetate. However, using Whatman No. 4 filter paper and after repeated freezing and thawing much esterase activity was unveiled (which is inhibited by eserine) in the reeler only. As has been suggested previously the paper effect may be due either to differential absorption of an inhibitor or to a dissociation of an inactive complex into active enzyme. A major portion of cholinesterase may be tightly bound to a particulate fraction and may be electrophoretically immobile since repeated freeze-thawing considerably enhances enzyme activity (probably by particulate decomposition). Cholinesterase is probably present in a cryptic form, becoming unmasked by the procedures employed; its relationship to the locomotor symptoms is worth investigating further.

Although the metabolic role of non-specific enzymes, e.g., esterases and phosphatases, is unknown, appearance of aberrant patterns in mice with genotypic differences (mutants) is noteworthy since they do not appear to be due to different reaction rates (zymograms were held for up to 36 hours at room temperature or at 4° C). It must be emphasized that the controls for the mutants are in some cases heterogeneous, e.g., for $d^l d^l$. Since $dse+/++du$ is heterozygous for d, not d^l, it may not be an appropriate control; in fact it is not even from the same strain and neither is the normal DBA/1J (d^+). Despite the heterozygosity of some of the controls for the mutants (not all stocks are alike, highly inbred or related) or the diversities in background, consistent differences between mutants and controls were observed. Studies on esterases (and also phosphatases) are continuing with the use of specific enzyme substrates and more critical evaluation of pH optima (Meier et al., 1962a).

In an analysis of the serum cholinesterase activities of three neurological mutants named tremulus (tr), waddler (wd), and quivering (qv) interesting findings were reported (Yoon, 1961). The enzymatic levels were lower in both male and female tremulus mice compared to their normal sibs. This was also true for quivering females, but no difference was found between quivering males and their normal brothers. On the other hand, waddler males showed a higher enzymatic level compared to their normal brothers, but no difference was found between waddler females and their normal sisters. In addition to the now well-established sex differences, some significant strain differences were observed. The final expression of these alterations in enzymatic levels is dependent upon the interaction of the original enzymatic level changes and the genetic constitution of the mice in which these changes occur. Serum cholinesterase in mice as shown by various tests is predominantly pseudocholinesterase; true cholinesterase activity is negligible if at all present.

Some mutants with well-developed clinical symptoms have been treated with a variety of neuropharmacological agents, e.g., tranquilizers, myorelaxants, and anti-convulsants. Although the number was small, certain mutants seemed measurably more sensitive in terms of sedation and toxicity of the drugs than did normal animals: muscle relaxants were effective for only temporary relief of spasms and tremors, and anti-convulsants such as monoamine oxidase inhibitors

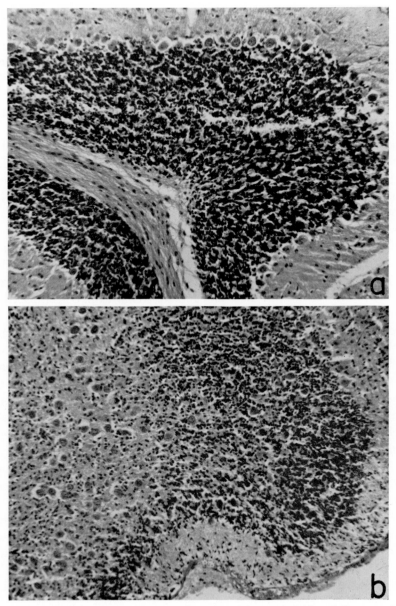

FIG. 4. Cerebellar cytoarchitecture of "reeler" [abnormal (b)] and littermate [normal (a)]. H and E. × 200. (From Meier and Hoag, 1962c.)

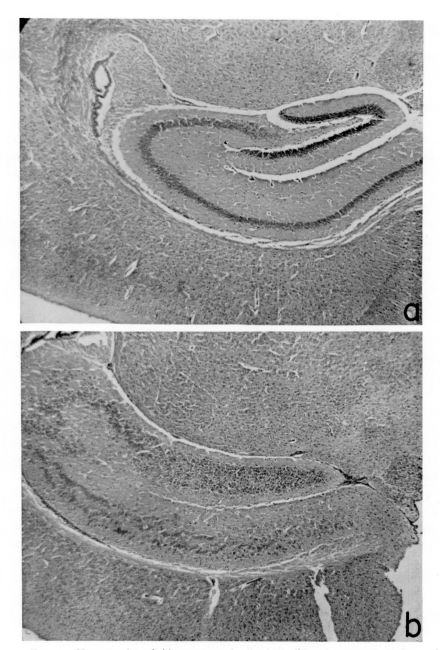

FIG. 5. Heterotrophy of hippocampus in "reeler" (b); the normal cytology of the hippocampus is shown for comparison (a). H and E. × 55. (From Meier and Hoag, 1962c.)

had no preventive or alleviative effect on convulsions associated with myelin degeneration (Rauch, 1960, personal communication).

Of interest are certain preliminary studies of Chai et al. (1962) on the spastic mouse (*spsp*). They found that aminoxyacetic acid abolished spastic systems while dilantin was somewhat effective; hydroxylamine and trimethadione were entirely without beneficial effects. The question of whether or not the effectiveness of aminoxyacetic acid lies in raising GABA levels in the spastic mouse is being investigated; hydroxylamine while it elevates GABA in most species does not have similar effects in mice (Roberts, 1961, personal communication). Roberts is currently involved in determinations of GABA levels in many different mutant types, comparing them with that in homozygous normal mice. Another fruitful endeavor would be to analyze generally the amino acid patterns of brains from mutants and controls, especially those with a homogeneous background.

c. Hereditary Muscular Dystrophy

Dystrophia muscularis (gene symbol *dy*) is a primary myopathy in mice that occurred as a spontaneous mutation in strain 129/Re (Michelson et al., 1955). It is transmitted as an autosomal recessive, and its expression is remarkably constant on a large number of genetic backgrounds (Loosli et al., 1961; E. S. Russell et al., 1962); this constancy of clinical manifestation in animals with divergent backgrounds provides a model for interpreting the three distinct human entities, each associated with a different mode of inheritance: (1) Duchenne type (sex-linked, recessive, boys); (2) limb-girdle (recessive autosomal, both sexes); and (3) facio-scapulo-humoral (dominant autosomal). The 129-*dydy* affects mainly the hind legs (Fig. 6), sometimes causing also head jerking, and is diagnosed at about 2 to 3 weeks (show microphthalmia and ichthiosis or stubby whiskers soon after birth, and alopecia at about 5 weeks of age); the incidence is about 19.1% per litter and lifespan is 105 ± 8.8 days). For comprehensive descriptions of the syndrome, reference is made to Michelson et al. (1955), Russell et al. (1962), and West and Murphy (1960). A certain interest in a possible interrelationship between mast cells and the etiology of hereditary muscular dystrophy in mice derived from the recent observation of an increase in number of mast cells in the skin of dystrophic mice

A. Pharmacologic Research in Inbred Mice

(O'Steen and Hrachovy, 1962). Accompanying this increased population of mast cells is a decrease in subcutaneous adipose tissue and an increase of collagenous connective tissue. Findings in the mouse will furnish basic information on the effect of an internal agent on the development and function of muscle tissue (whether or not it exactly corresponds to any of the human dystrophies). Already there is a bewildering array of evidence of deranged metabolism in dystrophic mice

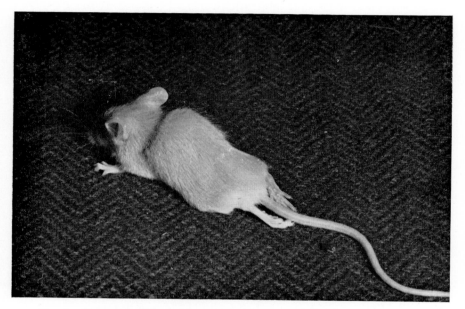

FIG. 6. Dystrophic (Dystrophia muscularis) mouse, $dydy$. (Courtesy of Photo- and Art Department, Roscoe B. Jackson Memorial Laboratory.)

(for review see Russell, 1961): There is abnormal creatine/creatinine balance, deranged lipid, carbohydrate, and protein metabolism and levels of a great many enzymes, e.g. abnormal distribution of serum aldolase, phosphohexoisomerase, lactic dehydrogenase, muscle aldolase, and phosphorylase (all glycolytic enzymes). Most biochemical studies on dystrophic mice have dealt with metabolic processes involved in glycolosis, protein breakdown, or oxidative steps; since no attention had been paid, until recently, to the alternate pathway of glucose metabolism, the pentose phosphate shunt, results are briefly mentioned.

Ribose-5-phosphate breakdown, heptose synthesis, and glucose-6-phosphate and 6-phosphogluconate dehydrogenase are increased, pentose phosphate isomerase is diminished (Canal and Frattola, 1962). As the pentose phosphate pathway is thought to be chiefly involved in supplying ribose-5-phosphate to DNA, it may be assumed that in dystrophic muscle this process proceeds at a faster rate; the nuclear proliferation observed by histological study may actually explain the increased need for DNA (see West and Murphy, 1960). Probably many of these alterations are the results rather than the causes of the basic effect; a complete list would be too voluminous. In order to determine primary effects the most useful approach is that of retrograde analysis, progressing toward earlier stages in an animal's life history, since the cause(s) of dystrophy must be shown to occur prior to the onset of clinical and histological manifestations. On this basis, elevated adenosine monophosphatase activity, accumulation of acetoacetate, and higher kidney glycine transamidinase activity which already appear at 2 weeks may be more closely related to the original gene action than other lesions occurring at later stages (Gould and Coleman, 1961, 1962).

Recently, two other abnormalities in metabolism have been found in muscle homogenates from dystrophic mice, one in the synthesis of pyruvate and one in its utilization (Coleman, 1962b). While muscle homogenates from normal mice synthesize pyruvate rapidly and accumulate a maximal amount in about 15 minutes, similar homogenates from dystrophic mice synthesize pyruvate considerably slower (40–50 minutes). Following the period of pyruvate accumulation, the rate of utilization in normal muscle exceeds the rate of production and the net result is a disappearance of pyruvate from the reaction mixture; again the rate of pyruvate disappearance in dystrophic muscle is much slower than in normal muscle homogenates. Studies on individual enzymes of pyruvate metabolism are now in progress in attempts to demonstrate the site of defective pyruvate utilization and also to relate the cause of the defect to that causing acetoacetate accumulation (Coleman, 1962b). Regarding acetoacetate accumulation, it was suggested that it may result from a failure of development in dystrophic muscle of the normal acetoacetate-metabolizing enzymes since it is possible to prevent its accumulation by the simultaneous incubation of normal and dystrophic muscle. Normal muscle, boiled prior to the incubation, is without effect, indicating that an enzyme rather than a heat-stable cofactor is responsible (Gould and Coleman, 1961). However, the ability of normal muscle may not consist of acetoacetate oxidation but rather of prevention of its formation. Also, whether or not acetoacetate accumulation is to be interpreted as an impairment of fat oxidation would require labeling of the endogenous lipid in dystrophic muscle. If this could indeed be shown, it would support the thesis that fat is a major endogenous substrate for skeletal muscle. No similar accumulation of liver acetoacetate has been found.

Perhaps a very fruitful approach to studying the crucial defect(s) in muscular dystrophy might concern an investigation of the pathways of intracellular hydrogen transport (for review see Boxer and Devlin, 1961). It is conceivable, for example, that the accumulation of acetoacetate results from lack of reduction by extramitachondrial $NADH_2$ (nicotinamide adenine dinucleotide, reduced form) to D(—)-β-hydroxybutyrate (acetoacetate-β-hydroxybutyrate "shuttle"). Preliminary experiments on $NADH_2$ oxidation by intact liver mitochondria from C57BL/6J, C3HeB/FeJ, and $B6D2F_1$ revealed that β-hydroxybutyrate added to the reaction mixture is not oxidized. Since α-ketoglutarate is not oxidized either, while succinate is, it may be suggested that there is an intramitochondrial deficiency of NAD (nicotinamide adenine dinucleotide) in mice. There is possibly an external path of oxidation of $NADH_2$ occurring in which β-hydroxybutyrate allows an external cytochrome c reductase to oxidize $NADH_2$ in a manner similar to that proposed by Boxer and Devlin for $NADH_2$ oxidation in the presence of cytochrome c in which the phosphorylating electron transport system is shunted. Possibly β-hydroxybutyrate acts as a mediator between $NADH_2$ and cytochrome a_3. These results are not yet verified, but perhaps the apparent difference between rat and mouse mitochondria may lie in the manner of preparation or the particular mitochondrial medium used (R. F. Baker, 1962, personal communication). The defect in pyruvate utilization could be explained by a relative inability of the mitochondria to oxidize pyruvate by way of the citric acid cycle. If, indeed, abnormalities in enzymatic hydrogen-carrying systems were demonstrated, the mutation could be shown to have affected the genetic information for formation and synthesis of some of the enzymes involved in intracellular hydrogen transport. Such a finding would also provide a model for biochemical investigation in cancer in which the absence of the enzymatic hydrogen-carrying systems is a factor in aerobic glycolysis (Boxer and Devlin, 1961). Transamidinase levels observed in heterozygotes always appear very close to the midpoint between levels of normal and mutant homozygotes. Excess dietary glycine stimulates enzyme activity and excess of creatine suppresses it (Coleman and Ashworth, 1959, 1960; Coleman, 1961). With reference to the increased monophosphatase, enzyme activity might result in a net decrease of available energy for metabolic maintenance (AMP is the starting point for ATP resynthesis); measurements of ATP should be made. An hypothesis has been presented (Tappel *et al.*, 1962) that increased lysosomal enzymes are the cause of hydrolytic and catabolic wasting processes. Obviously increased research emphasis should be placed on the last-mentioned metabolic alterations and those occurring early in life.

A recent most important development is the production of all-dystrophic litters of mice by artificial insemination (Wolfe and Southard, 1962); dystrophic juvenile or adult females were fertilized with sperm collected from dystrophic males. Obviously it is highly desirable to have available dystrophic tissue (for histologic and biochemical analysis) collected at prenatal and early postnatal stages when clinical diagnosis is not possible; also, since dystrophic males rarely breed it is impractical to use natural matings to obtain 100% dystrophic litters and there are no linked genetic characters known which could assist in early identification of dystrophic animals in segregating litters. It was found in a comparison of fertility and litter size of inbred and hybrid dystrophic mice that the latter were far superior to the former, providing a practical means to obtain all-dystrophic litters. Thus, the insemination procedure makes it possible to obtain dystrophic animals and tissues for earliest biochemical analyses and pharmacologic investigations.

d. Hereditary Anemias

Six different types of hereditary anemias are under investigation at the laboratory. Genes at three loci, dominant spotting, W, W^j, W^v (Grüneberg, 1939, 1942a,b; E. S. Russell, 1949; E. S. Russell and

A. Pharmacologic Research in Inbred Mice

Fondal, 1951; E. S. Russell *et al.,* 1957), Steel, *Sl, Sl*^{d} (Sarvella and Russell, 1956), and Hertwig's anemia, *anan* (Kunze, 1954), cause varying degrees of macrocytic anemia; homozygous flexed-anemia (*ff*) animals suffer from transitory, macrocytic, siderocytic anemia limited to fetal and neonatal life (Mixter and Hunt, 1933; Grüneberg, 1942a,b); animals homozygous for either jaundice (*jaja*) (Stevens *et al.,* 1958) or hemolytic anemia (*haha*) (Bernstein, 1959, unpublished), suffer from severe hemolytic disease with abnormal nucleated erythrocytes and extensive postnatal jaundice. Animals of five of the available genotypes die shortly after birth, and several other genotypes are semilethal; each anemia seems to present a different defect in hematopoiesis, and each is present before birth. Identification of the metabolic error in the primarily affected tissue is currently being pursued; also, measurements are being made of quantitative differences between effects of different alleles within the same series and between genes of different series. [To avoid difficulties from variations in genetic backgrounds on which these genes are segregating, development of congenic histocompatible strains differing only in the anemia-producing genes is important (Russell and Lawson, 1959).] In addition, since several important genotypes are available only in fetal and newborn stages, for the study of metabolism of fetal hematopoiesis micromethods are needed.

In studies on the action of W-series genes it was found that animals of severely affected genotypes are anemic because of a metabolic defect which either specifically delays the synthesis of protoporphyrin or nonspecifically arrests euthyroid cell maturation at a stage when synthesis of protoporphyrin is an important metabolic activity. The first evidence suggesting a biochemical basis for this arrest came from glycine-C^{14} incorporation experiments using anemic and normal littermates (Altman *et al.,* 1953; E. S. Russell, 1955); there was no difference between genotypes in time of erythrocyte appearance with labeled globin but in anemic mice appearance of cells with tagged protoporphyrin was greatly delayed. Similar results were obtained upon injection of δ-aminolevulinic acid, a protoporphyrin intermediate (E. S. Russell, 1955).

It may be expected that the responses of normal *ww* and anemic *WW*^{v} animals to radiation are extremely different; while a single dose of 200 r has little or no effect on the first, in the latter prolonged reduction of hematocrit levels ensues and in some cases even death (E. S. Russell et al., 1956; Bernstein unpubl., 1959). Also daily doses of purified erythropoietin induces increased hematocrit levels, marked reticulocytosis, and total blood volume in ww while WW^v are completely unaffected (E. S. Russell et al., 1959).

In addition to suffering from anemia, mice of all double-dominant W-genotypes lack pigment (100% white spotting) in their hair and are almost completely sterile (Mintz, 1957; Mintz and Russell, 1957); an interesting long-term consequence of the paucity of germ cells in W^vW^v mice is the development in all females of ovarian tumors (Russell and Fekete, 1958).

Homozygous $SlSl$ embryos and neonates (Bennett, 1956) and adult $Sl^d Sl^d$ show, similar to homozygotes of the W-series, a triad of pleiotropic effects in blood-forming tissue, germ cells, and hair pigmentation. Although the action(s) controlled by these loci involve processes specifically important in the metabolism of the three different tissues and are closely related, they may or may not influence different steps in the same synthetic pathways.

Still another anemia in the mouse [sex-linked recessive (proposed gene symbol sla)] was discovered by Falconer in 1958 at Edinburgh (Falconer and Isaacson, 1962). The blood picture characteristically consists of lowered red cell count (about ¾ of normal), decreased corpuscular hemoglobin, and reduced mean cell volume. As the mean cell diameter is more strongly reduced, it was concluded that the cells must be thicker than normal. There is also a reduction of hematopoietic tissue both in the liver and in the bone marrow. In spite of the smaller size of the anemic mice ($slasla$ females), their life expectancy is normal and the blood picture persists throughout the life span; females homozygous for the gene, and also hemizygous males, can be easily recognized at birth by their pale color. This difference disappears when they are a few days old, and therefore they have to be marked if they are to be recognized again; however, they may be persistently smaller than their littermates.

While the sex-linked anemia does not resemble any of the other inherited anemias in the mouse at all closely, it bears some resemblance to one type of hereditary human anemia, thalassemia or Cooley's anemia. Thalassemic homozygotes are clearly anemic at birth and remain so throughout life. Other points of resemblance between the two anemias are a proportionately greater reduction in the amount of hemoglobin and hematocrit than in the number of erythrocytes, variability of the size of erythrocytes (anisocytosis), abnormal shape of erythrocytes (poikilocytosis), the confinement of hemoglobin to the periphery of erythrocytes (pessary form), and an increase in the number of reticulocytes. Sex-linked anemia differs from thalassemia in that it has no effect in the heterozygote (thalassemic heterozygotes are slightly affected) and the fragility of the erythrocytes remains normal (it is often decreased in thalassemia). Also the bone marrow in sex-linked anemia is hypoplastic, whereas in thalassemia it is hyperplastic (Grewal, 1962).

e. Autoimmune Hemolytic Anemia

Bielschowsky et al. (1959; 1961, quoted from Holmes et al., 1961) have recently described an inbred strain of mice, NZB/BL, whose members spontaneously develop a high incidence of hemolytic anemia of an autoimmune type. The autoimmune nature of the disease has been confirmed by the finding that the affected mice develop a positive "direct" Coombs test; thus, while the study of autoimmune disease has been handicapped by the absence of any available condition in laboratory animals that was reasonably analogous to one of the human autoimmune diseases, a disorder closely similar to acquired hemolytic anemia of chronic type in man is now available (Holmes et al., 1961; Burnet, 1962). The disease can be successfully transmitted from Coombs-positive to Coombs-negative mice by means of spleen cells: a positive Coombs test, anemia, and other signs of blood destruction as well as splenomegaly are all induced.

In examining thymus sections from NZB/BL mice killed for transfer experiments, more than one half showed appearances which were interpreted as lymph follicles with germinal centers and having a close resemblance to the histologic findings in typical human cases of myesthenia gravis; there was severe thymic involution from age and stress, usually with complete atrophy of the cortex. Some of these showed an excess of lymphoid cells in the medulla, but the most striking feature was that in all mice more than 9 months of age in which the thymus cortex was still clearly shown, lymph follicles were present in the medulla (Burnet and Holmes, 1962). If, as has been claimed, autoimmune disease is a manifestation of the activity of lymphoid cells unresponsive to the normal homeostatic controls preventing development of cells reactive against body constituents, thymectomy should cure or alleviate the auto-immune disease. It is well known that in over 70% of human myasthenia gravis early thymectomy has this effect (Strauss et al., 1961). Experimental work based on this analogy is in progress (Burnet and Holmes, 1963, unpublished).

The present state of affairs seems to exclude the possibility that the Coombs-positive reaction is caused by the non-immunological liberation of a non-specific reaction globulin, but rather that an antibody is responsible. In view of the observation that the incidence of spontaneous lymphatic leukemia is low in NZB, reaction of the reticular tissue to the carcinogen 2-aminofluorene was tested. This compound significantly increased the occurrence of neoplastic lesions of reticular tissue in NZB. The lymphatic elements of NZB mice differ from those of the two related strains without hemolytic anemia, NZC and NZO, not only by their ability to form autoantibodies against red blood cells but also by a higher liability to undergo neoplastic transformation. Whether this is owing to a primary instability of these elements or to hyperplastic processes occurring in many lymphatic organs concomitant with the development of the hemolytic anemia remains to be investigated (Bielschowsky and Bielschowsky, 1962).

f. Diabetes Insipidus and Lesions in the Pituitary Gland

Water intake and urine output are above normal in mice of strain MA/J and MA/My. While the daily water consumption of a normal mouse varies between 5 and 8 ml, many of the MA mice drink over 40 ml daily. Water intake has been measured for parous females, virgin females, bred males, and virgin males, aged 3 to 18 months. It was found that average daily consumption was greater in bred than in virgin mice, that it increased with age in all groups, and that the increase was greatest and started earliest in parous females. In 25% of the parous females, 2% of the virgin females, 1% of the bred males, and none of the virgin males, the daily average was over 40 ml. Of

A. Pharmacologic Research in Inbred Mice

virgin males, 87% drank less than 10 ml daily, contrasted with 76% of bred males, 41% of virgin females, and 11% of parous females. Urine secretion was measured in the "heavy drinkers" and was found to be about equal in amount to water consumed, light in color, free of sugar, and of a low specific gravity. Since kidneys and adrenals are grossly and histologically normal, abnormal secretion of antidiuretic hormone is suspected (Hummel, 1961). Cysts and small alveolar glands are found in and near the posterior pituitary. The cysts, lined in part with ciliated epithelium, vary in size, are often multiple, and may obliterate large areas of the intermediate and neural lobes. The cysts appear to arise from enlargement of glands which are clustered at the junction of intermediate and neural lobes, and which are congenital and believed to be pharyngeal in origin. Cysts were found in 33 to 40% of male and female mice, aged 3 to 18 months, while alveolar glands were seen in 60 to 85%. Cysts and glands were also observed in neonatal mice. In the adult mice, about half of the lesions were classified as severe but the severity of the lesions could not be correlated with the excessive thirst characteristic of mice of these strains. The mice that drank the most water were not usually those with severe lesions. Small ciliated cysts, also of pharyngeal origin, were found in the anterior lobes of pituitaries with greater frequency than has been noted in mice of other strains (Hummel and Chapman, 1960–1961).

Strains MA/J and MA/My may be utilized for physiopharmacologic studies on the posterior pituitary; no work is yet in progress in this area.

g. Amyloidosis

Autopsies of inbred mice and their hybrids frequently reveal the presence of amyloidosis. Frequently associated with papillonephritis (the most common kidney disease of the mouse), amyloidosis occurs in several inbred strains as an inherited disease.

Now in progress are studies of strain A/Sn males to investigate the distribution of amyloid in various organs at different ages, the relationship of amyloidosis to kidney disease, the histochemistry of amyloid, and its possible association with lymphoid tissue reactions. No amyloid was noted in 1- to 3-month-old strain A/Sn males, but traces of this deposit were found in the lungs of 4- to 5-month-old animals. In mice, 6 months old and older, there was a gradual increase in the number of organs involved, as well as in the amount of amyloid present. By the 16th month all of the 16 organs surveyed contained amyloid, with the lungs and tongue the most prevalent sites of deposit. Papillonephritis was always observed at a time subsequent to the detection of amyloid. Except for a consistently negative iodine test, the staining characteristics of amyloid in the mouse were similar to those in man (West and Murphy, 1962).

There are several reasons for the usefulness of studying amyloidosis in inbred strains of mice: one is the need to search for unusual genetically determined protein variants and mechanisms whereby "usual" as well as "unusual" proteins become altered in disease. Recently a most intriguing hypothesis was proposed referring to fixation (but not initial stages of localization) of proteins, carbohydrates, and other materials in pathological deposits such as amyloid, hyalin (Benditt, 1961). Discussing the role of tanning in the pathogenesis of abnormal protein deposits, ceruloplasmin and catecholamines were cited as a potential tanning system (in analogy to tanning of the arthropod cuticle). The phenols and polyphenols (dihydroxyphenols as catecholamines, 3-hydroxytryptamine, 5-hydroxytryptamine, adrenaline, and noradrenaline) may be acted upon by ceruloplasmin (dehydrogenative coupling between ring structure) while the monohydroxyphenol compounds, tyramine and tyrosine, and the metahydroxy compound, resorcinol, are not oxidized. Another system (well known for its capacity to oxidize phenols to quinones in vertebrate skin) is tyrosinase; many protein components, e.g., fibrinogen and γ-globulin, have a propensity to precipitate under appropriate conditions and are known to have a substantial tyrosine content (also characteristic of proteins of the insect cuticle). Chemical and histochemical investigations are now under way to determine, in amyloid-susceptible and -resistant inbred mice, contents of components prerequisite for the tanning system (Meier *et al.*, 1962, unpublished).

h. *Amino-Acidemia and Amino-Aciduria*

Differences in the free amino acid (AA) content of tissues and fluids have been observed in human populations (normal and pathological) and in a variety of animal species. In mice, studies on AA's in normal and neoplastic tissue and in growing and regressing tumors have been reported (Roberts and Frankel, 1949; Roberts and Borges, 1955).

Statistically significant differences were disclosed in plasma between the genotypes in mice of the SEC/2 Gn-d strain for glutathione (*DD* high), glycine (*Dd* high), α-alanine (*DD* high), and valine-norvaline (measured together; *dd* high). No differences were found between the genotypes in the CBA/Ca-se strain. Between the genotypes segregating in the furless strain, there were significant differences found with respect to glutathione (*fsfs* high), aspartic acid (*fsfs* high), lysine (*Fsfs* high), and isoleucine-leucine (measured together; *FsFs* high). Significant differences in the level of arginine, lysine, valine-norvaline, and glutamic acid were found between the sexes within a particular strain. Male and female SEC/2 mice differed in the proportion of arginine in their blood, the males having the highest. CBA males differed from the females in their proportion of lysine (females higher than males), and valine-norvaline (males higher than females), while the proportion of glutamic acid in furless males exceeded that in females by a statistically significant amount. All of these sexual differences disappeared however, when all mice of all three strains were grouped according to sex and analyzed together. The strains differed statistically with respect to their relative proportions of glutathione, homocysteine, cystine, taurine, glycine, lysine, arginine, α-alanine, β-alanine, valine-norvaline, and isoleucine-leucine. Despite the presence of the many significant differences, a fairly standardized metabolic pattern could be generated for each inbred strain of mice (Hrubant, 1959).

Inherited amino-acidurias in man have received much attention and proved valuable in elucidating disorders of both metabolism and renal function. It is not unlikely that if mice (especially certain mutations) were screened, conditions biochemically similar to those known in man would be detected and become available for more intense study (Bennet, 1961). A mutation designated brain hernia (symbol *bh*) produces in homozygous condition abnormalities of skull, brain, and eyes, and polycystic kidneys 2 to 3 weeks after birth (Bennett, 1959). Results indicated no qualitative difference in patterns between *bhbh* and normal, but the amount of AA's excreted in *bhbh* was consistently abnormal. Proteinuria and generalized amino-aciduria in young *bhbh* occurred at a time prior to the development of severe polycystic disease;

after it becomes established, the urine becomes more dilute than normal. Similarly, quantitative but not qualitative differences have been found in dystrophic mice compared to normal mice.

Perhaps an even more important consideration than the free AA's in 129 mice is their excretion of urinary calculi: the rate at which calculi are excreted is a function of sex and of presence or absence of muscular dystrophy. The chemical make-up (which is markedly different from the organic matrix of crystallized human or bovine calculi) is a glycolipoprotein similar in a general way to that of the Bence-Jones protein with regard to the contents of phospholipid, cholesteral, and carbohydrate as well as high serine level but it deviates radically in the contents of other AA's. It seems possible that the 129 strain of mice may be a useful source for the study of calculus formation (Mc-Gaughey, 1961). The serine content of the total AA residue on a weight basis is approximately 15%; the glutamic acid content is also found to be high as compared with many proteins (about 20%); and the aspartic acid content is low (approximately 5%).

i. Hereditary Absence of Spleen

"Dominant hemimelia" (Dh) is a mutant gene of the house mouse causing striking skeletal and visceral abnormalities. While these abnormalities are similar in part to those of the group of "luxoid" mutants (skeletally, Dh heterozygotes are affected much the same way as luxate, lx, and urogenital abnormalities may occur as in luxoid, lu), the lesions are generally more severe. Aside from these defects all Dh hetero- and homozygotes lack the spleen; other lymphatic tissues such as the intestinal Payer's patches and the thymus seem unaffected. In addition, in homozygous mice the stomach and intestines are greatly reduced in size and length, the intestines usually ending blindly in the region of the colon; thus the lethality of these mice between 2 and 4 days of age is not surprising. Although Dh heterozygotes also tend to have a smaller stomach and shorter alimentary tract than normal ($++$) mice, those surviving into adult life show little or no reduction in viability or fertility.

The spleenlessness of the heterozygotes makes them useful subjects for radiobiological and physiological research. In an attempt to eval-

A. Pharmacologic Research in Inbred Mice 75

uate the immune responses of congenitally spleenless mice (*Dh* +), they were compared with normal animals (++): no distinct differences were obtained in antibody [CF (complement fixing) and hemaglutinin-] titers against *Pasteurella pneumotropica* 2 weeks following challenge; the experience with spleenectomized versus normal mice is similar. However, it was noted that heterozygotes consistently had lower serum proteins than did homozygous normal mice. Matings between the same parent heterozygotes produced two litters with the genotypes ++, *Dh*+, and *DhDh*; identification at birth is possible by the bright red spleen visible through the body wall of the normal and by the presence of curdled milk in the stomach of both the normal and the heterozygote. Measurements on the influence of spleenlessness, with (*Dh*+) and without intake of colostrum (*DhDh*), on blood proteins and of spleen plus colostrum showed marked differences in the pattern of blood proteins (total plasma proteins and absolute and relative levels of globulins) corresponding to the genotype; that of the *DhDh* represents a direct measure of the internal protein-synthesizing capabilities of mice born without spleens.*

Experiments underway are directed toward the specific function of the reticulo-endothelial system, e.g., alterations of protein (amino-acid) metabolism, leucocytic responses to various stimuli and phagocytic activities (Meier and Hoag, 1962e). The latter in particular may be subject to pharmacologic investigations.

j. Inherited Retinal Dystrophy

Retinal dystrophy is caused by the mutant gene "retinal degeneration," *rd,* which is present in several strains of mice. These include the three albino strains PL/J, SJL/J, and ST/J (Sidman, 1963, unpublished); recently, it was also discovered that CBA/J mice have no visual cells in the retina either (Green and Dickie, 1963, unpublished). Since offspring of CBA/J by C3HeB/FeJ, which carries the mutant,

---

* Rat fetuses (Sprague-Dawley) acquire most of the serum proteins present in the adult during the last part of gestation (Kelleher and Villee, 1962); recently, they also presented evidence of the production of serum albumin and at least some of the serum globulins during this period. Injection of C^{14}-labeled amino acids (leucine, methionine) intraperitoneally into rat fetuses in utero results in greater labeling of fetal serum proteins than when the amino acid is injected into the mother.

have the same abnormality, there is evidence that CBA/J has the genotype *rdrd*. In all of them the disease is inherited as a simple autosomal recessive (Lucas, 1958; Sorsby *et al.,* 1954; Tansley, 1954) and closely resembles retinitis pigmentosa, the most common inherited cause of human blindness.

Recent studies on C3H mice with retinal dystrophy both *in vivo* and *in vitro* (eye organ cultures) suggest the possibility that the disease represents an inability to use 11-*cis*-retinene. One reason for such failure would be lack of opsin, the protein that binds 11-*cis*-retinine to form rhodopsin; however, since a small outer sequent does form (and without opsin none should form), other or additional reasons for the inability to use 11-*cis*-retinene must be sought (Sidman, 1961). It may be theorized that the genetic defect in retinal dystrophy may be an error of protein synthesis manifesting itself in an abnormal opsin molecule. A rod outer segment may form with the abnormal opsin molecule as a structural component, but it breaks down since opsin in its abnormal configuration is unable to bind 11-*cis*-retinine.

It is of interest that the normal mouse eye *in vivo* converts all-*trans*-vitamin A to 11-*cis*-retinene by dehydrogenation and isomerization, while *in vitro* 11-*cis*-retinene is not formed. Yet the normal retina *in vitro* does make some outer segment upon addition to the medium of 11-*cis*-retinene. 11-*cis*-Retinene is, as part of rhodopsin, a structural component of the rod outer segment.

Anatomical changes that resemble both inherited retinal dystrophy and retinitis pigmentosa are obtained by placing rats on a vitamin A-deficient diet supplemented with an analog, vitamin A acid. The cell loss in miniature animals with inherited retinal dystrophy is, however, more rapid than the response of adult rodents to chronic vitamin A deficiency (Dowling and Wald, 1958, 1960). Sequential histological and histochemical analyses of eyes from C3H and C57BL/6 (normal) from birth on seem to indicate another possibility for retinal degeneration; namely, that a defect may also lie in the ground substance (mucopolysaccharides), deficiency of which causes the retinal cells to drop out (Meier, 1959, unpublished). A comparison between the retinas from 3-week-old C57BL/6 and C3H is illustrated (Fig. 7a and b).

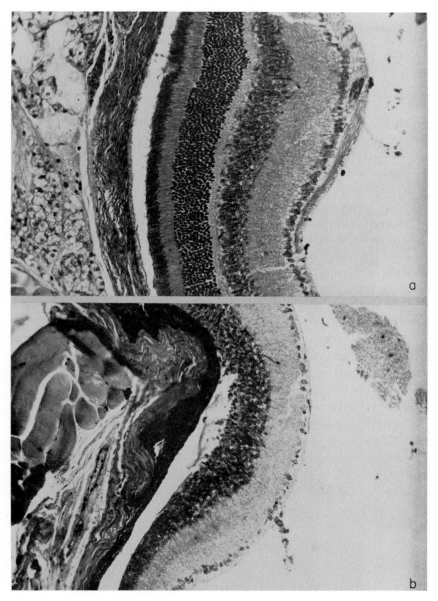

FIG. 7. Retina from a normal C37BL/6 mouse, 3 weeks of age (a) compared with that of an *rdrd* C3H (b). PAS stain. × 335.

k. Spontaneous Osteoarthropathy, Obesity, and Obstructive Genitourinary Disease

A multifold interest attaches to the STR/1N strain of mouse because of special characteristics that include spontaneous osteoarthropathy (degenerative joint disease), obesity while receiving a standard laboratory diet, a high incidence of hepatoma, low hemoglobin and high plasma lipids, primary polydipsia (see above), and an obstructive genitourinary disease affecting males only (Sokoloff and Barile, 1962).

Although obesity is associated with spontaneous osteoarthropathy, the increased load borne by the joints is not responsible for the development of degenerative joint (knees) disease. A genetic dissociation was demonstrated in $(A/LN \times STR/1N)F_1$-hybrids; while the offspring were almost as heavy as the STR/1N mice they had almost as little osteoarthritis as the (lean) A/LN mice. This demonstrates that the joint disease is under genetic control and its inheritance recessive; also, obesity is not the major influence in the development of the joint disease (Sokoloff *et al.*, 1960; Sokoloff, 1960). Chronic osteoarthropathy would lend itself to therapeutic approaches; very little has been done thus far.

A fairly constant observation at necropsy in male STR/1N mice before 16 months of age has been obstructive uropathy. Three forms have been observed: cystolithiasis, hemorrhagic occlusion of the urethral sinus, and suppurative vesico-urethritis. The lesions develop about hard plugs of altered seminal material impacted in the urethal bulb. The urinary sediment is of mixed type and apparently does not represent a single organic metabolite. Although microorganisms are consistently found in association with the cystolithiasis, it is uncertain whether they are the cause or result of the urinary status. The lesions are under genetic control, they do not occur in females and their development in males is prevented by castration.

B. Hereditary Characteristics of Pharmacologic Interest in Rats

Rats are used almost as widely as mice for procedures of drug evaluation (toxicity studies) and various biological assays (hormones). Maintenance of rat colonies for many years and use of rats for various

B. Hereditary Characteristics in Rats

laboratory investigations have resulted in a fair degree of standardization; also, through genetic studies a number of mutant genes have been identified and several linkage groups established. The following section considers briefly certain aspects of rat genetics.

1. GENETICS OF THE RAT

Castle (1947) concluded that the earliest attempts at domestication of the Norway rat (*Rattus norvegicus*) followed the discovery in wild populations of conspicuous mutants, albino, non-agouti blacks, and piebalds. These were captured and intercrossed resulting in the formation of a race of albinos homozygous for the three mutant genes, c (albino), a (non-agouti), and h (hooded). Rats of this genetic constitution were made the foundation stocks (around 1900) of a number of colonies, e.g., at the Wistar Institute in Philadelphia and the University of Chicago; most or all of the available strains were derived from them. In fact, "a survey of rat strains conducted in 1939 or thereabouts showed that all of the strains in this country, no matter what their name, carried some Wistar strain in them" (quotation of Dr. Farris; see Poiley, 1953).

a. Inbred Rat Strains

Unfortunately, the history concerning rat colonies is rather sketchy; Poiley (1953) provides some information on the "strains" maintained in the Animal Section of the National Institutes of Health (NIH). The "strains" include three albino groups, the Sprague-Dawley, the Holtzman-Rolfsmeyer, the Osborne-Mendel, and one pigmented group, the NIH black. All of these had been bred for many generations through the random selection of breeding stock and not via brother-sister matings; therefore, they do not conform with the definition of a strain (see above). However, although these animals have not been bred for specific genetic characteristics, they are distinct in a variety of reactions and also degrees of susceptibilities. Probably, because of these physiologic differences, detectable in laboratory experiments, the term strain is applied continually.

There are a few strains of rats that were developed by strict brother-sister inbreeding, e.g., the seven inbred rat strains maintained by the Inbred Nucleus of the Laboratory Aids Branch of the National Institutes of Health. These strains, together with their origin and certain characteristics, are listed in Table III.

TABLE III
INBRED RAT STRAINS[a]

Strain	Synonym	Origin and characteristics
A × C 9935	Irish	Inbreeding (b × s) started by Curtis and Dunning in 1926. To Heston 1945 at F30. To NIH at F40. Black agouti. Fertility good; ability to raise litters poor. Very nervous. High incidence of spontaneous tumors of neck and uterus. Kidney and genital tract abnormality in about 25%; this abnormality is unilateral (probably lethal if bilateral), kidney and ureter are cystic or absent and on the same side in females the ovarian capsule and/or uterine horn and/or proximal portion of the uterine horn may be absent and in males on the side with the kidney abnormality the vas deferens and/or vesicular gland may be missing
BUFFALO	None	Inbreeding (b × s) started by Heston 1936 in non-inbred Buffalo stock obtained from Morris. To NIH 1950 at F10. Fertility medium; ability to raise litters medium. Docile. Albino
FISCHER 344	None	Inbreeding (b × s) started by Curtis and Dunning in 1920. To Heston 1949 at F51. To NIH 1950 at F51. Fertility good; ability to raise young good. Docile. Albino
M-520	Marshall 520	Inbreeding (b × s) started by Curtis and Dunning in 1920. To Heston 1945 at F41. To NIH 1950 at F54. Fertility good; ability to raise young good. Docile. Albino
O'GRADY	None	Inbreeding (b × s) started by O'Grady. To Heston in 1948 at F28. To NIH 1950 at F32. Fertility poor to medium; ability to raise young medium. Nervous. Albino
O-M	Osborne-Mendel	Inbreeding (b × s) started by Heston in 1946 from non-inbred Osborne-Mendel stock obtained from J. White. To NIH 1950 at F10. Fertility medium to good. Ability to raise young medium to good. Docile. Albino
WISTAR	None	Inbreeding (b × s) started by Heston in 1942 from non-inbred Wistar stock obtained from Nettleship. To NIH 1950 at F15. Fertility poor; ability to raise young fair. Docile. Albino

[a] From Jay (1953).

B. Hereditary Characteristics in Rats

b. Mutations of the Rat

In the course of domestication and maintenance of rat colonies the total number of mutations increased to 23 in 1947; the kinds of mutant genes that have been recognized over the years and their linkage relation are presented in Table IV (Castle, 1947). Certain of these are of

TABLE IV
MUTATIONS OF THE RAT[a]

Designation	Genetic symbol	Linkage group	Nature
Albino	c	I	Absence of pigment from coat and eyes
Non-agouti	a		Absence of wild coat pattern, uniform black
Hooded	h		White except head and back stripe
Pink-eyed yellow	p	I	Coat yellow, eyes pink
Red-eyed yellow	r	I	Coat yellow, eyes red
Curly	Cu	II	Hairs of coat and vibrissae curved
Brown	b	II	Black pigment of coat and eyes replaced by brown
Stub	st	IV	Short stubby tail
Ruby-eyed dilute	c^d	I	Allele of albino gene, c, pigmentation diminished
Curly$_2$	Cu_2		Coat hairs and vibrissae strongly curved
Kinky	k	IV	Coat hairs and vibrissae strongly curved
Lethal	l	I	Skeleton imperfect
Blue	d		Black pigment diluted (clumped) to yield a blue
Hairless	hr	III	Hair lost at about 4 weeks of age
Wobbly	wo	III	Ataxic locomotion
Waltzing	w	I	Runs in circles
Incisorless	in	II	Incisors lacking
Anemia	an	II	Young anemic at birth, lack of red blood cells
Cataract	Ca		Opaque lens visible in unpigmented eyes, pink-eyed or albino
Jaundice	j		Skin and hair yellow at birth and later
Shaggy	Sh	II	Hair and vibrissae curved, closely linked to curly
Silver	s		Black coat interspersed with white hairs
Fawn	f		Dilutes black to tawny, blue to fawn

[a] From Castle (1947).

considerable biomedical interest, e.g., jaundice (j) and also a recently detected mutation causing extreme obesity (fa). The potentialities of many other mutant genes such as those affecting the nervous system (wobbly, wo; and waltzing, w), and those causing anemia (an) or hairlessness (hr), have not been realized as yet.

c. *Physiologic Properties of Strains and Mutants*

Data on the physiologic properties of rat strains and mutants are scarce; because of possible interest, certain properties will be listed. Morris *et al.* (1953) reported on a study involving high and low efficiency of food utilization of two strains of rats; Luecke *et al.* (1945), from the same laboratory (University of Minnesota), found that high efficiency of food utilization appeared to be accompanied by a low requirement for thiamine and a high requirement for riboflavin. Heston (1938) described how on a rickets-producing diet (either low in P or Ca), rats of the red-eyed yellow (Castle) strain showed a bent nose, while rats of Hoppert's colony did not; both strains showed a normal nose on a normal diet. Different frequencies of duodenal ulcers from pantothenic acid deficiency in adult rats occurred in the 9B strain (10%) and 13C strain (80%). Other differences relate to growth rates, basal metabolism, oxygen consumption, etc. (for review see Zucker, 1953).

Ershoff (cited by Emorson, 1953) subjected rats of the Sprague-Dawley, USC, and Long-Evans strains to low temperature and made the following observations: less than 5% of the Sprague-Dawley rats survived a temperature of 3°C for a period of 2 weeks; 50% of the USC strain survived 2 weeks and more than 75% of the Long-Evans strain survived for this period.

In feeding three different supplements to a hairless mutant, described by Roberts (1924/1925 and 1926/1927), Martin and Gardner (1935) found that (1) cysteine caused growth of a hair coat that persisted for a month or more, (2) cystine produced a flash of hair that disappeared in a week or two, and (3) glutathione had no effect on hair growth at all. It was concluded that the genetic defect might lie in the splitting off, from a larger compound, of cystine or in the reduction of cystine to cysteine. The two dwarf mutants described by Lambert and Sciuchetti (1935) and Wooley (1948) might show en-

B. Hereditary Characteristics in Rats

docrine abnormalities similar to those of the dwarf mouse; comparative studies have not been made. The subsequent sections specifically consider hereditary characteristics of pharmacologic interest.

2. Strain Differences in Drug Action

A practical example of strain differences in drug action relates to toxicity studies of *thiourea*; the LD_{50} in wild Norway rats was found to be 1340 ± 230 mg/kg, in Harvard (Long-Evans) rats, 44 ± 13 mg/kg, and in "Hopkins" rats, 4.0 ± 0.2 mg/kg. Obviously, as a rodenticide, in sparing wild rats but readily killing laboratory rats, thiourea does not represent a very promising poison (Dieke and Richter, 1945). In studies on Chloretone (trichlorobutanol), the stimulation of ascorbic acid and glucuronic acid excretion was greater in Wistar than in Sherman rats (Mosbach et al., 1950).

Also, the finding of Brodie (1952, quoted by Williams, 1959) should be mentioned: among eight lines of rats capable of metabolizing *antipyrine* there were two that inactivated the drug very slowly while the other six dealt with it very rapidly.

A fair amount of work concerned strain differences in free-choice consumption of ethyl *alcohol*; evidence has been produced that differences in alcohol choice may be an index of emotionality (Adamson and Black, 1959) and also that alcohol choice is a useful character for genetic analysis. Reed (1951) and McClearn and Rogers (1959) showed variations between strain of rats (similar to those in mice) in voluntary consumption of alcohol. Myers also found marked differences in alcohol preference between G-4 rats* (high) and Wistar rats; while emotionality may cause this difference the possibility also exists that the effect may result from differences in alcohol metabolism in the two strains or from a differential resistance to the intoxicating effects of a given concentration of alcohol in the blood. It is noteworthy that alcohol selection in these experiments was clearly related to environmental temperature, suggesting that there may be other mechanisms involved that are yet to be uncovered.

---

* G-4 rats are an inbred strain (well over 20 generations of brother-sister matings) maintained at the Jackson Laboratory and were descended from a cross between two laboratory stocks by Grüneberg in 1937 (Grüneberg, 1939); they are non-albinos and carry the recessive "pink-eye dilution" gene.

Activities of hepatic enzymes have been studied in female Sprague-Dawley rats killed after 3, 6, and 12 weeks of chronic alcohol (ethanol) intoxication (Figueroa and Klotz, 1962). Alterations were found to vary with the length of time of intoxication but were due to the direct toxic effects of ethanol on the liver cells and only in part to reduced food intake. Alcohol dehydrogenase reacted initially with an elevation and later, following 12 weeks of intoxication, a decrease (independent of caloric or food intake). Alcohol dehydrogenase activity increased in those rats that received sucrose for 6 weeks; the fructose molecule of sucrose may play a part in this elevation. Isocitric dehydrogenase was decreased on a restricted food intake, alcohol further enhancing this reduction. The transaminases were also affected in the alcohol-intoxicated rats with a period of decreasing glutamic-pyruvic transaminase activity and a moderate glutamic-oxalacetic transaminase elevation; no similar changes occurred in isocaloric and weight control animals.

3. PHARMACOLOGIC EFFECTS OF SEROTONIN AND RESERPINE IN RATS

Serotonin (5-HT) decreases the body temperature of both rats and mice (Hoffman, 1958; Lessin and Parkes, 1957); also it lowers the blood pressure in rats (Erspamer, 1954). Both effects are enhanced following adrenalectomy. It is unlikely that the increased toxicity is caused by an impaired degradation of 5-HT since the activities of both monoamine oxidase (Brodie *et al.*, 1959) and ceruloplasmin (oxidase) are unchanged following adrenalectomy (Blaschko, 1960). It may be that the increased sensitivity to 5-HT is owing to a lack of compensatory mechanisms (Garattini *et al.*, 1961c).

In intact rats (Sprague-Dawley strain) a remarkable increase in 5-HT toxicity is observed with age while it was stated that in mice age is of little influence; comparison of the Sprague-Dawley and Vister (Wistar descended) strains revealed no difference in LD_{50}15: Sprague-Dawley 198 mg/kg (172.2–227.7) and Vister 195 mg/kg (163.9–232.1), respectively, of 5-HT intravenously (Garattini, 1962, personal communication). However, there is a significant strain difference upon adrenalectomy: Sprague-Dawley 9.2 mg/kg (7.2–11.6) and Vister 27.5 mg/kg (13.8–32.7); Long-Evans rats are intermediate, 14.5 mg/kg (9.9–22.6). Garattini (1962, personal communication) observed for

B. Hereditary Characteristics in Rats

monoamine oxidase activities in various tissues of adrenalectomized rats the values given in Table V.

TABLE V
Monoamine Oxidase Activity in Tissues of Adrenalectomized Rats[a]
(LD_{50} in mg/Kg)

Tissue	Sprague-Dawley	Vister
Liver	161 ± 34	141 ± 16
Kidney	57 ± 10	45 ± 10
Brain	29 ± 15	38 ± 19
Lung	104 ± 25	110 ± 20

[a] Garattini (1962), personal communication.

Similarly, strain differences occur upon intraperitoneal injection of reserpine, SU 3118 (syrosingopine), and SU 5171 [methyl-18-O-(3-N,N-dimethylaminobenzoyl)reserpate] into adrenalectomized rats (Table VI).

TABLE VI
Drug Toxicities in Strains of Rats Following Adrenalectomy[a]
(LD_{50} in mg/Kg)

Compound	Sprague-Dawley	Vister
Reserpine	0.13(0.102–0.166)	0.38(0.288–0.502)
SU 3118	1.85(1.58–2.16)	2.56(2.00–3.28)
SU 5171	1.40(1.19–1.65)	1.16(0.95–1.42)

[a] Garattini (1962), personal communication.

4. Strain Differences in Cholinesterase Activity and Acetylcholine Concentration of the Brain

Selection experiments have been concerned with a variety of traits in many different organisms. In rats, an example relates to high and low response of the ovary to a standard injection of gonadotropin (Kyle and Chapman, 1953). Recently, successful attempts were made to modify by selection the activity of enzymes. Roderick (1960) developed two strains differing in cortical and subcortical cholinesterase by a selective breeding program using the cortical cholinesterase of the sire as the basis; since the rats were derived from two stocks, Dempster and Castle at the University of California, four subgroups were obtained labeled RDH (Roderick Dempster high cholinesterase), RDL, RCH, and RCL (Roderick Castle low cholinesterase) strains. A number of correlated responses to selection were found: (1) brain size negatively associated with cholinesterase; (2) brain density positively associated with cholinesterase; (3) brain weight correlated with body weight, both of which were negatively associated with cholinesterase and subcortical cholinesterase; and (4) subcortical cholinesterase positively associated with cholinesterase.

Bennet *et al.* (1958b) published results indicating individual and strain differences in cortical and subcortical cholinesterase activity in two strains of rats that were developed through selective breeding for maze brightness (S_1 strain) and maze dullness (S_3 strain); the S_1 was consistently higher than the S_3 in brain cholinesterase activity. These observations were interpreted as indicating a relationship between adaptive behavior and brain cholinesterase activity (Rosenzweig *et al.*, 1958). This relationship tended to be explained by the part which the acetylcholine-cholinesterase system plays in transmission at synapses in the central nervous system. In answer to the question as to whether or not cholinesterase and acetylcholine concentrations are genetically linked, results obtained by Bennet *et al.* (1960) indicate that the genetic mechanisms controlling substrate concentration and enzyme activity are independent. Acetylcholine-cholinesterase ratios have not yet been obtained; however if different ratios occur, it would suggest that neither acetylcholine nor cholinesterase alone is an adequate index of efficiency at cholinergic synapses. Furthermore choline acetylase may be intimately involved and determinations of it made. This should allow for testing the hypothesis that the efficiency of neuronal functioning and adequacy of adaptive behavior are functions of the balance among choline acetylase, acetylcholine, and cholinesterase.

Measurements of lactic dehydrogenase support the hypothesis that the observed correlations between cortical cholinesterase activity and adaptive behavior is not a reflection of the general metabolic rate in the brain but rather is specific to the acetylcholine-cholinesterase cycle. Values obtained for the two enzymes, cholinesterase and lactic dehydrogenase, differ in the following respects: (1) cholinesterase activity shows a more sharply differentiated pattern of regional distribution within cortex and subcortex than lactic dehydrogenase; (2) cholinesterase clearly rises from 30 to 100 days between the strains while lactic dehydrogenase does not; (3) cholinesterase is significantly different between the strains, lactic dehydrogenase is not; (4) cholinesterase

B. Hereditary Characteristics in Rats

values are significantly correlated from locus to locus within the brain, lactic dehydrogenase values are independent of locus; and (5) both cholinesterase and lactic dehydrogenase activities show some correlation with each other in the cortex but not in the subcortex. While these findings may be interpreted to mean that cholinesterase activity is not simply an index to general metabolic level of the brain, the differences between cholinesterase and lactic dehydrogenase as well as the differences in cholinesterase between the strains may be related to variations in architectonic structure of the cortex (Bennet et al., 1958a).

5. Sex-Linked Differences in Drug Action

Aside from strain differences there are numerous compounds whose pharmacologic effects are sex-dependent. A sex difference was reported for *carisoprodol* (N-isopropyl-2-methyl-2-propyl-1,3-propanediol dicarbamate) in Sprague-Dawley rats, but it could only be observed in adult animals (also mice and guinea pigs) and not in immature rats; the duration of paralysis produced by carisoprodol (as determined by the length of loss of the righting reflex at 19°–22° C) was about 2.3 times longer in adult females than in males (Kato et al., 1961). Since brain and serum concentrations at the end of paralysis were higher in females than in males, the differential effects of the drug may be inferred to be due to sex differences in the metabolic capacities rather than to differences in drug sensitivity. The anabolic action of testosterone appears to be an important factor in accelerating carisoprodol metabolism. Although the pathways of carisoprodol breakdown are not yet known exactly, there exist interesting analogies to that of *pentobarbital*: (1) a sex difference is present; (2) enzyme activities (located in liver microsomes, O_2 and $NADPH_2$ requiring) are increased upon pretreatment with phenobarbital or phenaglycodol [2-(p-chlorophenyl)-3-methyl-2,3-butanediol]; and (3) action is inhibited by SKF 525 A (β-diethylaminoethyldiphenylpropylacetate hydrochloride).

Sex differences in the pharmacologic effects similar to pentobarbital and carisoprodol have also been found for *strychnine, picrotoxin,* and *octamethyl pyrophosphoramide* (OMPA), but again only in adult (Sprague-Dawley) rats (Kato et al., 1962a). Also, the high activities of drug-metabolizing enzymes may be due to the anabolic action of testosterone.

Analogies to increasing activities of drug-metabolizing enzymes with age also occur regarding the activities of some hepatic microsomal $NADPH_2$-dependent steroid-metabolizing enzymes (Kato et al., 1962b). No activity is found in rat fetuses (19–20 days); very low activity occurs in newborn, but enzyme activity increases progressively after birth.

6. CORTISONE RESISTANCE IN PREGNANT RATS

Among the biochemical alterations that ensue in rats after the 15th day of pregnancy are a lowered alanine-glutamic transaminase activity, lowering blood amino nitrogen and blood sugar, and increased liver weight and nitrogen retention. In view of the well-known protein catabolic effects of cortisone, administration of this hormone to pregnant rats was expected to reverse or at least impair this anabolism with a consequent adverse effect on gestation. By this procedure, abortion or macerated fetuses and congenital abnormalities can be produced in rabbits and mice (Courrier and Cologne, 1951; DeCosta and Abelman, 1952; Fainstat, 1954). However, in the rat (Wistar strain), doses of cortisone as high as 25 mg/day produced no significant effect on the course of pregnancy. This marked biochemical resistance to cortisone was not altered upon removal of the fetuses, but was abolished when both fetuses and placentae were removed. It is to be concluded that the placenta is essential for this resistance, but resistance to cortisone is not part of an over-all resistance to protein-catabolic compounds since pregnant rats exhibit no resistance to glucagon (Curry and Beaton, 1958). The resistance of pregnant rats to cortisone provides a model for studies of the causes of a similar resistance of pregnant women to this hormone (Assali and Suyemoto, 1954; Margulis et al., 1954).

7. CHOLESTEROL SYNTHESIS IN RAT BRAIN AND THE PROBLEM OF MYELINATION

Cholesterol synthesis in the mammalian brain takes place only during the phase of myelination (Srore et al., 1950). It is still present and measurable in rats several days after birth (Grossi et al., 1958). Comparative studies on the pathways of cholesterol synthesis in liver and brain of Long-Evans rats (12 days old) revealed the following differences: (1) in the liver, mevalonolactone is efficiently incorporated

B. Hereditary Characteristics in Rats

while in the brain it is poorly utilized (Tavormina et al., 1956; Garattini et al., 1959) and (2) the lack of efficient mevalonolactone incorporation by brain (as well as intestine, testis, and skin) is apparently due to its greater utilization of acetic acid (Fumagalli et al., 1962, personal communication); these observations therefore would explain the quantitative differences both *in vitro* and *in vivo* of acetic acid-1-C^{14} or 2-C^{14} and mevalonolactone-2-C^{14} into cholesterol.

Recently it has been found that in brain slices or homogenates of 12-day-old Long-Evans rats the incorporation of potassium mevalonate-2-C^{14} is several times greater than that of mevalonate-2-C^{14}, probably due to more rapid penetration; in contrast no difference was observed in livers or, if anything, incorporation of potassium mevalonate was less than mevalonolactone (Fumagalli et al., 1962, personal communication). While these data suggest that mevalonolactone is an unsuitable substrate for brain cholesterol synthesis, it does not exclude the possibility of its being formed during endogenous synthesis. Also, it is possible that the equilibrium between laconate and the acid form regulates the rate of mevalonic incorporation into brain cholesterol. The presence of a brain lactonase has not yet been demonstrated.

Little information concerning the sequence of myelination in the mouse has been published. Folch et al. (1959), using classic histological techniques for demonstrating myelin, observed that there was virtually no myelin present until the end of the first week after birth. The degree and extent of myelination then increased until it reached apparent completion at 50 days of age. Uzman and Rumley (1958) also state that mouse brain has very little myelin up to 10 days of age. Luce (1959), by means of electron microscopy, found evidence of beginning myelin deposition about axons in the spinal cord of a one-day-old mouse. From these findings it would seem that in the mouse more time is available than in the rat for studies of myelination. Also, reference has already been made to certain "staggering mutants" and the dilute-lethal mouse mutant in which myelin degeneration proceeds simultaneously with myelination. Degenerating myelin is observed in dilute-lethal mice ($d^l d^l$) in the vestibulo-spinal, spino-cerebellar, and fecto-spinal systems and appears within a day or two after the first signs of myelination; normal littermates of dilute-lethal mice (Dd^l, and DD) as well as control mice from other strains do not exhibit degenerating myelin. It would be interesting to study brain cholesterol synthesis in these three genotypes.

8. HYPERCHOLESTEROLEMIA IN SUCKLING RATS

Serum cholesterol levels of suckling rats are at least twice that of adult rats; for example the average in Sprague-Dawleys at 13 days is 232 ± 12 mg/100 ml, while at 40 days the average is 98 ± 3 mg/100 ml and in Long-Evans rats at 13 days, it is 221 ± 15 mg/100 ml and at 40 days it is 96 ± 15 mg/100 ml (Bizzi et al., 1962, personal communication). This high level of serum cholesterol is probably related to the high fat concentration in milk and the ready absorption of fat by the intestine of newborn rats. The cholesterol level rapidly decreases after weaning while prolongation of the sucking period maintains the elevated level. Contrary to induced hypercholesterolemia by feeding cholesterol, there is comparatively less free cholesterol than in weaned rats. Liver cholesterol biosynthesis is not increased in suckling overweaned rats, and the levels of cholesterol in liver and aorta are also comparable. It may be, however, that there is an impairment of cholesterol disposition (decreased catabolism and excretion) that is responsible for the hypercholesterolemia; also, aside from an elevation of cholesterol, desmosterol and phospholipids (not triglycerides) are increased in suckling rats. Of particular interest is the fact that hypercholesterolemia of suckling rats may be used for the screening and evaluation of potential hypocholesterolemic drugs; thyroxine and its congeners reduce serum cholesterol at doses that are ineffective in adult rats. Similarly difenesemic acid and benzmalacene decrease serum cholesterol. Without effects were heparinoid, biphenylbutyric acid, nicotinic acid, triparanol (MER-29), hexestrol, and ethionine; the last three exert hypocholesteremic properties in adult rats (Bizzi et al., 1962, personal communication).

Apparently no relationship exists between the level of lipid biosynthesis and the actual lipid content of a given tissue; for instance, interference of cholesterol biosynthesis does not directly influence the level of cholesterol in blood or tissues of Long-Evans rats (Garattini et al., 1961a). Examples relate particularly to biphenylacetic acid and ethionine; a comparable inhibition of cholesterol synthesis of the liver may not alter blood cholesterol levels (biphenylacetic acid) or decrease

B. Hereditary Characteristics in Rats

it markedly (ethionine). More information is needed regarding the mode of action of drugs influencing cholesterol levels. Control of cholesterol biosynthesis in the mouse has been the subject of several investigations. It is known that cholesterol itself lowers the rate of hepatic cholesterol synthesis, but appears to have little effect on extrahepatic cholesterol synthesis; also, bile acids have been shown to control the rate of hepatic cholesterol metabolism through a feedback mechanism (Beher and Baker, 1959). Intracellularly accumulating cholesterol reduces the rate of cholesterol synthesis and bile acids, by decreasing the acetylation rate from CoA, and also reduces the amount of substrate available for conversion. In various studies on the rate of cholesterol biosynthesis it was found that cholesterol biosynthesis is controlled by cholic acid and its conjugates—by a double feedback reaction (Beher *et al.*, 1962).

9. HEREDITARY OBESITY ASSOCIATED WITH HYPERLIPEMIA AND HYPERCHOLESTEROLEMIA IN THE RAT

Zucker and Zucker (1961) described a new mutation, designated "fatty," which appeared spontaneously in the 13M rat stock of their laboratory. The condition is due to a single recessive gene for which the symbol fa had been chosen (the normal allele is Fa).

Obesity becomes evident as early as 3 weeks of age, is very apparent at 5 weeks, and the average weight at 40 weeks is from 600 to 800 gm; normal littermates of the same age weigh between 295 to 480 gm. Male "fatties" have usually normal appearing sex organs although they are only occasionally fertile, and the females show a small, underdeveloped uterus and are uniformly sterile. Therefore, phenotypically normal heterozygotes ($Fafa$) of both sexes must be depended upon for continuing the stock.

Aside from the obesity, the most striking sign of "fatty" is lactescence (milky appearance) of the blood serum. This hyperlipemia (estimated total lipids 2730 mg/100 ml versus 367 mg/100 ml in controls, $fa+$) involves a 10-fold rise in total fatty acids (6.00–8.81 millimoles versus 0.68–0.83 millimole in normal controls) and a 4-fold increase in cholesterol (218–489 mg/100 ml versus 89–95 mg/100 ml of controls) and lipid phosphorus (20.3–30.8 mg/100 ml versus 6.0–7.7 mg/100 ml in controls).

Apparently there is a striking error of fat metabolism; both the great excess of fat (probably low density lipoproteins) in the serum and the excessive fat deposition in the tissues occur on a low fat diet or even under dietary restrictions (Zucker and Zucker, 1962). The NEFA (non-esterified fatty acid) level characteristic of fed "fatties" of any age was consistently about twice that for fed normals, and was much like that of fasted normals; this observation suggests that tissues from obese rats are certainly capable of liberating NEFA and in fact, at rates comparable to normal. Although there are no significant differences between homozygote normals (*FaFa*), the heterozygotes show a tendency toward higher weight, increased fatty acids and cholesterol.

While livers are moderately fatty, frequently also renal calculi with uni- or bilateral hydronephrosis occur in older animals (a year or more). It is noteworthy that in obese rats of one year or older, aortas have so far shown no signs of atheromatous plaques despite the severe hypercholesterolemia (Zucker and Zucker, 1961). Although "Yale" rats have been described as a case of "hereditary obesity" (Mayer, 1955), probably because they are distinctly larger than Wistar rats, their total carcass fat content is 20% (or less) which is less than one-third to one-half of the levels (40–50%) in *fafa* (Harned and Cole, 1939; Pickens *et al.*, 1940). Hereditary obesity (*fafa*) differs from experimentally induced obesity either by surgically produced hypothalamic lesions (Brobeck *et al.*, 1943) or feeding of certain high fat diets (Mickelsen *et al.*, 1955) in that the "deciding factor" is not simply excessive food intake (hyperphagia); hyperphagia, even if marked, only moderately raises serum fatty acids and cholesterol. It also differs from the obese mice (*obob*) in its lack of hyperglycemia, but is similar in that both are caused by a basic metabolic disturbance* (see above). However, in view of the role of insulin in obesity, the finding of enlarged to greatly hyperthrophied pancreases in obese rats requires further exploration; in the O-H mice there is increased islet tissue without gross hypertrophy of the pancreas.

It is clear that, although hereditary obesity of rats has not yet been fully investigated, a new and most valuable tool for metabolic and

---

* There are also obese mice of the SH silver strain (Salcedo and Stetten, 1943) and an obese New Zealand strain (Bielschowsky and Bielschowsky, 1956) which differ physiologically and metabolically from the O-H mice.

B. Hereditary Characteristics in Rats

pharmacologic studies of obesity is at hand; it represents one more channel of investigation in addition to the hereditary conditions of mice, including the O-H (*obob*), adipose (*adad*; Falconer and Isaacson, 1959), and heterozygotes of yellow (A^ya; Grüneberg, 1952a). Obviously, in the obese rats, particular importance attaches to the severe hyperlipemia.

10. Hereditary Non-hemolytic Jaundice

While there appear to be several different heritable defects of conjugation in man, the Crigler-Najjar syndrome is best understood from the viewpoints of heredity (autosomal recessive), primary defect (deficiency of glucuronyl transferase), and clinical aspects (gross bilirubinemia, unconjugated type, usually kernicterus and persistent jaundice).

An apparently identical condition occurred in a strain of Wistar rats maintained at the Connaught Laboratories (Gunn, 1938). The inheritance pattern is that of a single recessive gene. A marked lag in growth was found to be characteristic for the majority of homozygous mutants; associated with stunting were wobbly gait or partial paralysis; offspring from homozygous matings were jaundiced at birth, while those from hybrid matings developed jaundice within the first 12 hours after birth. Heterozygous animals appeared free of clinical signs.

In studies on the primary defect, several investigators (e.g., Carbone and Grodsky, 1957) found the affected rats to suffer from a deficiency of glucuronyl transferase; heterozygous rats were assayed to have slightly subnormal transferase activity while bilirubin levels were normal (Johnson *et al.*, 1959). Treatments of these jaundiced rats with 3,4-benzpyrene had little or no effect on the formation of *o*-aminophenol glucuronide (Inscoe and Axelrod, 1960); however, the liver microsomes from these benzpyrene-treated animals showed a 7-fold increase in the ability to hydroxylate acetanilide. The neurological signs associated with hereditary jaundice are due to kernicterus which is analogous to that seen in infants and which was intensified by various sulfa drugs (e.g., sulfadiazine, sulfamethoxine, sulfisoxazole) and salicylates (Johnson *et al.*, 1957). The paradoxical lowering of serum bilirubin associated with intensified staining of the brain cells provided evidence for *in vivo* competition between bilirubin and the drugs for binding sites on serum albumin.

Recently, inhibition of glucuronosyl transferase by steroids was observed. This steroid inhibition may play a role in the jaundice of the newborn and also in the jaundice that follows the administration of certain anabolic steroids (Dowben and Hsia, 1962). The rate of conjugation of uridine diphosphoglucuronic acid (UDPGA) with an aglycone by guinea pig microsomes *in vitro* was studied. The aglycones used were *o*-aminophenol, *p*-nitrophenol, and 4-methylumbelli-ferone. Pregnanediol ($4.2 \times 10^{-5} M$) resulted in 50% inhibition. Δ^1-17α-methyltestosterone, testosterone, pregnanediol glucuronide, 17α-methyltestosterone 6α-methyl-17α-hydroxyprogesterone, and 17α-ethyl-19-nortestosterone also inhibited the conjugation, while cholesterol, pregnanedione, pregnenolone, 17α-hydroxyprogesterone, cortisone, estradiol, estrone, progesterone and methylandrostenediol did not inhibit. While several steroid glucuronides were found to inhibit the conjugation of an aglycone with UDPGA, they were less strong inhibitors than free steroids. Kinetic studies with pregnanediol glucuronide and 17α-ethyl-19-nortestosterone (norethandrolone) showed the inhibition to be of the competitive type.

C. Pharmacologic Studies in Hamsters

The term hamster is applied to several different species of animals; although over sixty-six varieties of subspecies are known in the four genera of true hamsters, only three have been used to any extent in laboratory studies. Two of these, the golden or Syrian hamster (*Mesocricetus auratus*) and the Chinese hamster (*Cricetulus griseus*) will be considered.

The *golden hamster,* since its first introduction as laboratory animal, has been used in the study of leishmaniasis and other experimentally produced diseases ranging from parasitic infections, bacterial and viral diseases to dental caries and cancer; more recently it had been found that it accepts both homologous and heterologous tissue grafts (normal and tumorous) without or with only little preconditioning. Aside from certain morphologic characteristics (e.g., lymphatic system, cheek pouches) the golden hamster is of particular interest because of

C. Pharmacologic Studies in Hamsters

its physiologic (biochemical) peculiarities in steroidogenesis, a marked sexual dimorphism in reaction to estrogens, and the occurrence in an inbred strain of primary hereditary polymyopathy. Other than hereditary muscular dystrophy a number of mutations have occurred most of which affect pigmentation. Only one such is mentioned (mutant allele symbol c^d) because of the occurrence of counterparts in rabbits, mice (Himalayan), and cats (Siamese). The adult mutant possesses pink (pigmentless) eyes and a white pelage but with dark (pigmented) ear pinna. The young hamster at 21 days is devoid of skin melanism and has flesh-pink ears; after 31 days the ear color darkens. These observations strongly suggest that the phenotype is that of a thermosensitive acromelanic albino (Robinson, 1957).

The *Chinese hamster* is the most recent (for a review, see Yerganian, 1958) rodent adapted to laboratory conditions. This species was thought to be of value because of several unique morphological features, parasitic relations, and disease susceptibilities. In more recent years particular interest has centered on the fact that its chromosome number with $2n=22$ is the lowest yet to be observed among eutherian laboratory mammals which generally have 40 or more in the diploid ($2n$) complement (e.g., golden hamster $2n=44$). Extensive studies on chromosomes have dealt with subjects pertaining to radiation cytogenetics, tumor cytology, action of chemicals (alkylating agents) on mitosis and chromosomes, etc. Initiation of several docile inbred lines has provided tissues with which to start *in vitro* analyses. Of particular importance was the occurrence during inbreeding of a spontaneous hereditary diabetes mellitus (see Section III,C,4).

1. Amphenone in the Golden Hamster

Depending on the size, number, and distribution of lipid droplets in the cortical cells, and the presence or absence of a positive Schultz test (cholesterol), mammalian adrenal cortices are either lipid-rich or lipid-poor structures; most common laboratory animals qualify as having lipid-rich adrenals excepting the golden hamster, *Mesocricetus auratus*. Aside from histochemical differences the hamster as compared to rats is also unusual in that its adrenal gland exhibits an inverted sexual dimorphism, the absolute and relative weights of the male glands being greater; this difference is abolished by gonadectomy

(Kupperman and Greenblatt, 1947). Adrenalectomized hamsters are better maintained by progesterone than deoxycorticosterone acetate (Snyder and Wyman, 1951). These morphologic and physiologic peculiarities suggest that there may be basic functional biochemical pathways between animals with lipid-poor and lipid-rich adrenals. Amphenone-B [3,3-bis(p-aminophenyl)-butanone-2-dihydrochloride], a drug known to interfere in steroidogenic pathways, did not produce the marked adrenomegaly or increase in adrenal cholesterol content seen in other species; neither was there an effect on corticoidogenesis as determined by urinary steroid excretion. However, it results in the excretion of a metabolite that interferes with the Zimmermann test for 17-ketosteroids. Since there occurred no piling up of cholesterol in the hamster adrenal, and consequently no adrenal hypertrophy, a difference in the metabolic pathways for steroid synthesis had to be inferred. Hamsters may not utilize cholesterol as an adrenal steroid precursor, but may condense small metabolic units directly to a C-21 rather than a C-27 compound (cholesterol); some of the metabolic steps required for amphenone inhibition may be lacking. This postulate should be supported by the use of labeled substrates; if verified, direct incorporation of labeled acetate into adrenal corticoids without labeling of the adrenal cholesterol pool (which had been demonstrated in cell-free preparations of hog adrenal) may be the predominant process in the intact adrenal gland of the hamster (Marks *et al.,* 1958).

2. Malignant Renal Tumors in Estrogen-Treated Male Golden Hamsters

Chronic estrinization of male and female golden hamsters shows a variety of effects after $3\frac{1}{2}$ to $8\frac{1}{2}$ months. These effects include growth retardation, decrease in bone formation, and changes in the reproductive system and endocrine organs (Koneff *et al.,* 1946). In all males treated with estrogen (20 mg pellets of diethylstilbestrol) for 250 days or longer, except those in which diethylstilbestrol-cholesterol pellets were implanted, renal tumors develop; they always occur bilaterally and are cortical in location. The tumors are usually adenomatous, may become ultimately malignant, but do not metastasize. No tumors occurred in any females or castrated males (Kirkman and Bacon, 1950).

C. Pharmacologic Studies in Hamsters

3. Polymyopathy of Hereditary Type in an Inbred Strain of Golden Hamster

A generalized polymyopathy and myocardial necrosis occurs in all individuals of an inbred strain of hamsters developed at the Bioresearch Institute (Cambridge, Massachusetts) and identified as BIO 1.50 (Whitney). The myopathy appears after 3 weeks of age and progresses to about 220 days at which time deaths occur from massive myocardial necrosis. This disease is the third hereditary muscular dystrophy in animals (Homburger et al., 1962). Muscular dystrophy of mice differs from that of hamsters in that it occurs extremely early in life (2 weeks), interferes with reproduction and kills in a short time (3–6 months); hamster dystrophy is also different from a hereditary myopathy of chickens (Julian and Asmundson, 1960) with respect to distribution and types of histologic lesions as well as onset of the condition. However, since muscular dystrophy in man is now conceded by all to be a hereditary disease (Adams et al., 1962), the discoveries of similar conditions in animals should greatly stimulate experimental work on muscle diseases. Preliminary biochemical investigations on hereditary polymyopathy has also revealed differences as compared to muscular dystrophy of mice (Homburger, 1962, personal communication); future findings will be pertinent to possible pharmacologic approaches. While it is of interest from a comparative pathological viewpoint that in many respects the disease is similar to certain sheep with scrapie (Bosanquet et al., 1956) and necrotizing myopathy of man (Denny-Brown, 1960), the occurrence of venous thrombi, consisting of "diseased" skeletal muscle fragments, is puzzling (Homburger and Baker, 1962, personal communication); whether or not there is any similarity to a metastatic process is undetermined.

4. Spontaneous Hereditary Diabetes Mellitus in the Chinese Hamster (*Cricetulus griseus*)

Diabetes mellitus was recognized, in 1958, in a random colony of Chinese hamsters (*Cricetulus griseus*) maintained at the Children's Cancer Research Foundation, Boston, and arose spontaneously during the course of inbreeding (Meier and Yerganian, 1959). At the time, several sublines of the major families were approaching the fourth generation of continuous brother-sister mating. Following establishment of a pathological diagnosis, which included alterations of beta-cells (Figs. 8a and b), hypoinsulinism (Zahnd *et al.,* 1960, personal communication) as well as diffuse vascular (especially glomerular) changes (Fig. 9), hyperglycemia and ketonuria, an extensive breeding program was inaugurated to evaluate the genetic background. Clear

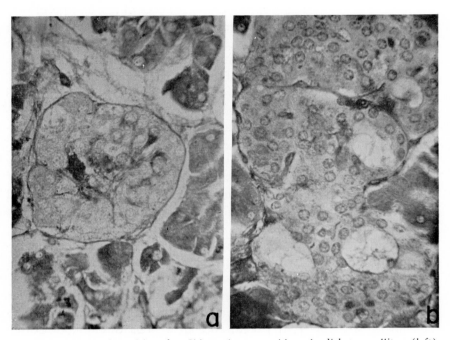

FIG. 8. Langerhans islet of a Chinese hamster with early diabetes mellitus (left), and a hamster with severe diabetes of about a month's duration, showing large intracellular vacuoles and strands containing PAS-positive material (right). Aldehyde–fuchsin. × 304. (From Meier and Yerganian, 1959.)

evidence is now at hand indicating that the disease is hereditary, and transmitted as an autosomal recessive trait: inbreeding from the fourth to the twelfth brother-sister generation had raised the incidence of diabetes to 100% of the offspring; random bred animals having 50%

FIG. 9. Ischemic and hypocellular glomerulus from a diabetic hamster (above, H and E, × 150); intercapillary glomerulosclerosis (below, PAS, × 150). Both sections were obtained from the same animal. (From Meier and Yerganian, 1959.)

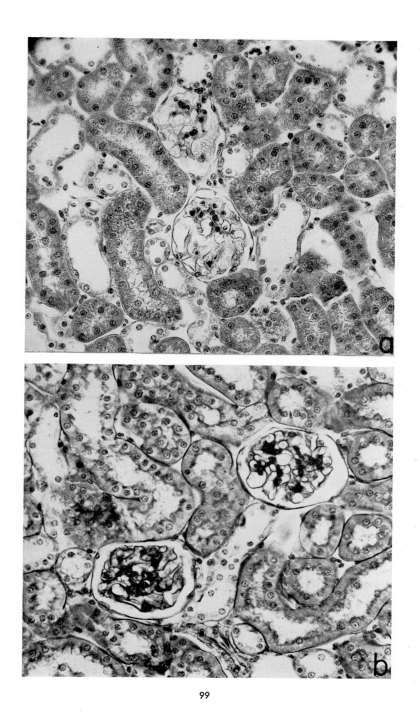

homozygosity failed to show diabetes, while the degree of penetrance (or intensity of suppression) was enhanced when the over-all genetic background reached 65%, or higher, homozygosity. Since penetrance or severity of the disease varied among families or inbred lines, the over-all genetic constitution governs the time and degree to which the metabolic disturbance is expressed. The clinical onset of diabetes may occur as early as 18 days or as late as 250 days. Accordingly, and relative to sexual maturity (which occurs at about 45 to 60 days in the Chinese hamster), it was possible to distinguish arbitrarily 3 groups of diabetics, the "juvenile" groups showing symptoms within the first month of life, the "adolescent" group, 1 to 2 months of age, and the "adult" group including animals diagnosed after sexual maturity. Littermates among inbred lines generally display similar levels of glucose (e.g., 0.5–2%, or higher, glucosuria) and ketone bodies, and they respond equally to therapy; this feature permits replication of trials or multiple therapy on both sexes. In the beginning, maintenance of fertility of diabetic parents was an ever-present obstacle; also, there were considerable losses during pregnancy (abortions, resorption) and among live-born animals (deaths shortly after delivery, cannibalism). Many hamsters were totally dependent upon careful chemotherapeutic control of diabetes for continued existence and maintenance of fertility (Meier and Yerganian, 1960).

Obviously, the discovery of a diabetic state, genetically transmitted, in the Chinese hamster not only allows studies on inheritance of diabetes or screening of potentially hypoglycemic agents, but offers in addition the feasibility of systematic study of "prediabetes," the effects of pregnancy, and the presence of associated pathologic states. Thus, insight may be gained into the origin of specific lesions observed in diabetes mellitus, i.e., the pancreatic islet cells, angio-, retino-, and neuropathies, and their occurrence relative to time and manifestation of metabolic alterations. Interesting observations have recently been reported on severe peridontal involvement in diabetic hamsters corresponding closely to the experience in man with diabetes mellitus, and characterized by pocket formation, inflammation, and bone resorption (Cohen et al., 1961). Like diabetes, spontaneous adenocarcinoma of the pancreas has been encountered rarely among laboratory rodents in contrast to a significant statistical association of this malignancy

C. Pharmacologic Studies in Hamsters

with diabetes in man; since pancreatic adenocarcinoma was among the first of the spontaneous tumors recorded for Chinese hamsters, its association with a diabetes-susceptible strain may not be merely a chance possibility (Poel and Yerganian, 1961). Also, several years ago, groups of hamsters, including some from lines in which diabetes mellitus appeared, had been sent to the Chester Beatty Institute in London; a number of these developed spontaneous pituitary tumors.*

a. Toxicity Studies in Normal Hamsters

Phenformin (N^1-β-phenethylformamidinyliminourea hydrochloride) was given orally (by stomach tube), subcutaneously, and, intraperitoneally. Tolbutamide, 1-butyl-3-(p-tolylsulfonyl)urea (sodium-Orinase) was administered both subcutaneously and intraperitoneally. NPH (aqueous) insulin was used by the subcutaneous route. All hamsters received single injections either at one or two sites. When compound solutions intended for gastric tube feedings exceeded 1.5 ml, they were administered in two divided doses within 15 minutes. The animals were maintained on Purina Laboratory chow with supplementary wheat germ flakes and water ad libitum.

A total of 116 normal adult hamsters were employed to establish maximum tolerated doses of each compound and to evaluate drug

---

* A syndrome consisting of obesity, due to marked accumulation of adipose tissue beneath the skin of the breast and abdomen as well as in the peritoneum and mesentery, and hyperglycemia is associated with primary or transplanted pituitary tumors of *parakeets*. Pituitary tumors are the most common tumors of shell parakeets (*Melopsittacus undulatus*). While in the birds with primary tumors (obesity thought to be due to pressure on the hypothalamus), this mechanism is unlikely because of the appearance of obesity in birds with subcutaneous transplants. It appears that the level of somatotropic hormone (STH) activity found in assays of pituitary transplants may produce the obesity and hyperglycemia in the parakeet. The fact that marked hyperglycemia was not encountered in birds with relatively smaller primary pituitary tumors suggests that STH is present in low concentration. Growth hormone has not previously been identified in the pituitary of normal birds; it is obviously present, but whether or not it is identical with STH of mammals is undetermined.

While pancreatectomy in the carnivorous horned owl produces diabetes, in grain-eating birds such as the chicken, duck, and pigeon pancreatectomy does not result in any significant disturbance of the blood sugar. It may be that in these birds this is due to the loss of alpha cells (glucagon) at the time of pancreatectomy. As a corollary, the loss of degranulation of beta cells and the presence of large numbers of alpha cells in parakeets with pituitary tumor transplants may be linked with hyperglycemia (for a review see Schlumberger, 1956).

toxicity. Two to four animals (males and females) were used for each dose level of the drugs and each route of administration.

Evaluation of drug toxicity was difficult because many animals died of hypoglycemia before any toxic effects could be observed histologically. Clinical evidence of toxicity, however, occurred with both phenformin and tolbutamide, depending upon the route of inoculation, even before hypoglycemic shock resulted. The interval between drug administration and death, from either hypoglycemia or other causes, was timed routinely.

b. Therapy in Diabetic Hamsters

In diabetic hamsters, comparative studies were done on the relative effectiveness of the agents in reducing elevated blood sugar levels. The least toxic route of administration was chosen for therapy.

In mildly diabetic hamsters (blood glucose between about 150 to 300 mg/100 ml), it was possible to reduce elevated blood sugar levels to normal by appropriate doses of each of the drugs. The correct dosage was reached by gradual increases in daily amount. The starting dose depended upon the degree of diabetes, but was roughly one which lowered the blood glucose level in both healthy animals to about half normal. Normal blood sugar levels in severely diabetic animals, excreting 10 to 15% or more urinary sugar per diem, could be maintained only with NPH insulin; both phenformin and tolbutamide were ineffective. In a few instances, diabetic hamsters died soon after the administration of any of the three compounds, before a normal sugar level could be reached. Some animals required increasingly higher drug levels over a period of weeks, indicative either of resistance or progressive worsening of the condition.

Since some deaths were attributed to the daily handling and transfer of animals from one cage to another for breeding purposes, it was decided to supply phenformin or tolbutamide in the drinking water. This proved helpful and time saving as the number of diabetic hamsters increased. Only animals totally dependent upon insulin were injected daily. Phenformin was given at 30 mg/100 ml, and tolbutamide at 134 mg/100 ml in water bottles. The bottles were renewed daily. The dilutions were more or less arbitrary, but daily water intake

C. Pharmacologic Studies in Hamsters

and relative effectiveness of the compounds were taken into consideration. Most animals maintained themselves at moderately hyperglycemic levels, the mortality rate was considerably reduced from about 30 to 5% over a half-year period, and fertility and the number of complete pregnancies increased. Quantitative data on survival are not yet at hand. In the course of therapy it became obvious that the sensitivity of diabetic animals to hypoglycemic drugs was greater than that of normal animals; comparatively less compound was required to lower elevated sugar levels than to induce hypoglycemia in normal hamsters. For instance, 2 to 40 units (U) of NPH insulin per hamster (30–33 gm) would (often when 70 to 100 U or more was required) reduce elevated blood sugar levels to normal, even in severe diabetics; in order to induce hypoglycemic shock in normal controls 40 to 90 U were required.

In mild to moderately severe cases of diabetes, there was an almost linear relationship between reduction of the urine sugar excretion and dosage of NPH insulin. In severely diabetic hamsters, insulin was the only effective drug. It is therefore doubted, contrary to the reports of others, that phenformin can act in complete absence of insulin. It was estimated that phenformin replaced roughly one-fourth of the insulin required for adequate therapy in mild to severely diabetic hamsters. Diabetic animals were, within limits, comparatively more sensitive to the hypoglycemic action of phenformin than normal animals.

Although LD_{50}'s had not been determined in other animal species at the time of Ungar's report in 1957, semiquantitative data on the action of phenformin on the blood sugar in various normal animals would suggest that the species most sensitive to subcutaneous inoculation of phenformin is the rhesus monkey. The species most resistant was the rat. Normal Chinese hamsters were approximately ten times more resistant than the rat and fifty times more resistant than the rhesus monkeys to phenformin given subcutaneously. This observation was made by comparing the LD_{100}'s of our normal hamsters with those determined and published for other animals. The subcutaneous administration, however, proved somewhat erratic, since many animals died almost immediately following injection, or within 15 minutes thereafter. The oral route of administration caused fewest toxic deaths.

Tolbutamide, administered subcutaneously and intraperitoneally, was about as effective in normal hamsters as was oral or intravenous administration in normal rats.

The doses of subcutaneous NPH insulin (about 500 U/kg) required to lower the blood sugar levels to about half of normal were larger for normal hamsters than for any other species.

From the data presented in this report and those available in the literature it may be extrapolated that on a weight basis, normal hamsters were approximately ten to twelve times more resistant to tolbutamide, eight times more resistant to phenformin, and about 150 times more resistant to NPH insulin, than was man. The exact reasons for these and the underlying metabolic species differences are obscure (Meier and Yerganian, 1960b).

c. Aspects of the Prediabetic State

Since there has in the past few years accumulated a fair body of evidence suggesting that patients with diabetes mellitus have a demonstratable derangement of the basement membrane of many, if not all, of the blood vessels of the body; this vascular disease, being perhaps the major cause of illness and death in diabetes, may precede the hyperglycemic manifestations of the diabetes; in fact it may be responsible for the failure of the islet cells to elaborate adequate amounts of insulin. Should these formations be correct, the concepts of pathophysiology of diabetes might be altered radically; also the approaches to therapy may become modified. In the following, four observations in Chinese hamsters will be discussed that pertain to the prediabetic state; the third and fourth being particularly pertinent to the concept outlined above. The first relates to observations in the *pancreatic islets* in offspring of diabetic parents (Meier and Yerganian, 1960a). Sequential comparisons with increasing age of islet numbers, sizes, and alpha:beta cell ratios disclosed striking differences between offspring of diabetic, prediabetic,* and normal parents. The great majority of diabetics' offspring revealed islets which, characteristically, were more numerous, larger (due to hyperplasia), or both. Islets regarded as hyperplastic consisted almost entirely of beta cells. Formation of islets

---

* "Prediabetic" mating refers to crosses between parents prior to their becoming clinically diabetic. Since the age of onset of diabetes is fixed in each line and predictable, brother-sister matings could be made while diabetes was still unapparent.

C. Pharmacologic Studies in Hamsters

appeared to occur from small ducts, the islets forming cuffs around them or incorporating them. At 2 hours of age, proliferation as indicated by the presence of mitotic figures seemed more pronounced in the diabetic's offspring than those of prediabetic or normal parental mating. At one day of age, the islets in progeny from diabetic parents were about twice as large (approximately 100 μ.) as in normal controls (50 μ.) and were about four times as numerous.

Cell counts per islet (counting all cells in the plane of the largest diameter) averaged 120 cells in offspring of diabetic animals and forty in those of normals; alpha:beta cell ratios, as obtained in sections stained with aldehyde fuchsin, were found to be 1:7 (diabetic) and 1:5 (normal), respectively. The beta cells were finely granulated and the granulation was evenly distributed throughout their cytoplasm. Islets of offspring from matings of prediabetic animals showed less difference when compared with normals. At 5 to 7 days, comparisons of the islets in the progeny of diabetic and normal animals disclosed smaller differences relative to number, size, and alpha:beta ratios, and at 10 to 14 days they were almost completely abolished. The diameter of the islands measured roughly 75 to 120 μ and alpha:beta cell ratios were 1:4, as in normal adult animals. While the islets were still proliferating as evidenced by mitotic activity, their growth occurred between 5 and 14 days (more rapidly in the normal progeny), mitotic indexes being about three to five per islet against one or two in the prediabetic progeny. The beta cell granulation had become abundant, essentially filling the entire cell.

No increase in the size of the islets was noted at 23 days or later, except in one family of diabetic parents (VSY). This strain was completely inbred, all progeny revealing clinical diabetes at 18 days of age. At 14 days, secondary islet cell hyperplasia occurred, associated with beta cell degranulation; a few cells showed a hydropic change, also. In other families, degranulation took place gradually, requiring weeks; the islets were either almost totally degranulated or had only a few well-granulated cells left. Large intracellular vacuoles were noted, with strands crossing them and staining, in part, periodic acid-Schiff (PAS)-positive. This material has been identified as glycogen by both additional histochemical procedures and electron microscopy. In man, among the less probable causes islet hyperplasia has usually been considered as being due to maternal hyperglycemia, the degree reflecting the severity of the mother's diabetes. Other evidence cited favoring this hypothesis stemmed from experimental work, e.g., the induction of islet cell hyperplasia in rats in response to hyperglycemia. It seems noteworthy, however, that in the Chinese hamster some degree of islet hyperplasia occurred in offspring of parents with normal blood sugars (hypoglycemic therapy; and from "prediabetic" matings). The mechanism underlying islet hyperplasia, therefore, is still obscure. Further studies are needed to determine the relationship of the degree of hyperplasia and the severity of parental diabetes, the influence of the prediabetic state, and therapy. Soon after birth, in the normal offspring rapid islet proliferation ensues, abolishing differences existing between them and the prediabetic progeny in the first few days to one week of life. The period of apparent normalcy prior to the onset of degenerative changes (i.e., hydropic degeneration, deposition of PAS-positive material, leading to an eventual beta cell deficiency) varied greatly with each line.

The second observation concerns the presence of *hydronephrosis* in one of the diabetic lines, detectable as early as 2 to 3 days after birth (Meier and Yerganian, 1959). In the absence of any evidence of mechanical urinary obstruction and compression of the renal parenchyma, it is to be considered as another genetically determined expression. Since it has generally been stated that in man congenital anomalies are more common in infants of diabetic mothers, the findings lend support to this possibility.

The third observation pertains to microscopic findings in the kidney (Meier and Yerganian, 1959; Lawe, 1962): "plastic" changes (nodular-hyaline material) are not seen in diabetic hamster kidneys; neither is anything resembling an exudative lesion. Diffuse *glomerular disease,* although less specific than the nodular type in human diabetic nephropathy, is present in the hamster and consists of widening of the basement membrane and of an increase in PAS-positive material lying over and spreading (intercapillary) beyond the boundaries of the basement membrane. Strikingly, there is severe degree of ischemia and hypocellularity. These changes differ from those associated with a diffuse intercapillary glomerulosclerosis as an expression of age rather than of a specific disease; the extent of the lesion seemingly correlates

C. Pharmacologic Studies in Hamsters

with both time of onset and severity of diabetes, e.g., in one line of hamsters with diabetes characteristically manifest at day eighteen of life, intercapillary glomerulosclerosis develops parallel or perhaps even prior to and in progression with the diabetic condition as judged by careful histologic examination of the pancreatic islands and glucosuria. The most severe degree of glomerular change may be seen at one month of age, many animals dying as a result of uremia. Age most certainly is of inconsequential influence. Also, loss of glomerular function is greater than in any physiologic aging process and disproportionate in time if, in view of the shortened life span of diabetic hamsters, diabetes would only induce premature aging. These observations therefore argue for a specific diabetic lesion. It is of interest that in the obese-hyperglycemic mouse, *obob,* renal glomerular changes do not occur even if the hyperglycemia persists for months (Meier, 1963, unpublished); therefore, the suggestion that the renal disease involving the glomerular basement membranes is a primary lesion may be enhanced.

Fourth, the point to be made is that the renal lesion (and perhaps other angiopathies) may be a primary one and not necessarily consequential to beta cell degeneration. Electrophoretic patterns of *serum proteins* in families with a high incidence of spontaneous diabetes reveal alpha-2 levels to be 2 to 3 times the normal values (Green *et al.,* 1960). In normal animals, the alpha-2 proteins generally are between 5 to 10% of the total serum proteins (Fig. 10a and b). When high-incidence families were randomly hybridized by single or double crosses involving two or four grandparents, respectively, of diabetic background, normal values of alpha-2 are re-established.

In man, slightly elevated alpha-2 serum proteins, and protein-bound carbohydrates in patients with diabetes have also been reported. This elevation of alpha-2 values occurred generally when the diabetes was complicated by vascular and other secondary changes. In the Chinese hamster, however, alpha-2 serum proteins are increased, even prior to the onset of clinical diabetes and microscopic evidence of intercapillary glomerulosclerosis. Breeding experiments have been performed, using elevated alpha-2 serum proteins as a genetic marker in an attempt to increase the incidence of spontaneous diabetes. The extremely high levels of alpha-2 (up to 35%) observed in diabetic hamsters may be

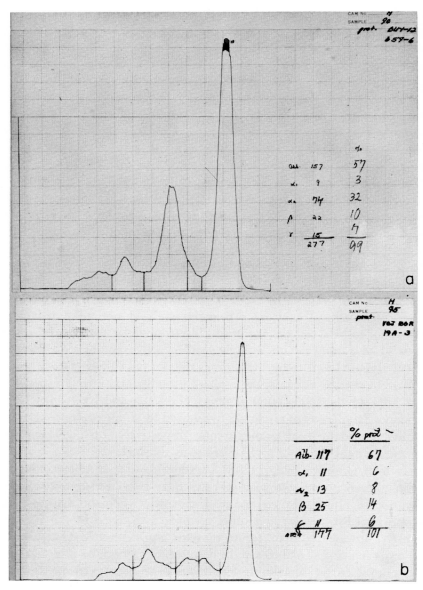

FIG. 10. Elevated alpha-2 globulin in a Chinese hamster from a family with a high incidence of spontaneous diabetes mellitus (a); an electrophoretic pattern from a normal hamster (b). (Courtesy of M. N. Green; for reference see Green et al., 1960.)

due either to an increase of normally existing alpha-2 proteins, or perhaps to new and different molecules which are a genetic reflection of the prediabetic state. Chemical analysis of the alpha-2 globulin fractions in diabetic and non-diabetic hamsters would be required in order to distinguish between these two possibilities. The fact that in obese-hyperglycemic mice there is no elevation of alpha-2 globulins provides further evidence to indicate that the alpha-2 elevation in the Chinese hamster is a primary lesion and perhaps the earliest detectable (Meier, 1963, unpublished).

Obviously, the genetic aberration as occurring in certain strains of Chinese hamsters affects many tissues of the body (pleiotropic effect): changes prior to beta-cell decompensation and onset of hypoinsulinism with clinical manifestations of the disease consisting of specific elevation of alpha-2 serum globulins, degrees of islet hyperplasia, other genetic expressions, and angiopathy. These and other well-established findings, e.g., the high incidence of fetal complications of pregnancy (both in women and hamsters), support the existence of a systemic metabolic aberration long before the insulogenic mechanism becomes overwhelmed.

D. Pharmacologic Responses Controlled by Heredity in the Rabbit

The rabbit has been contributing valuable genetic material for the solution of a wide range of genetic, medical, and other biological problems. Perhaps most significant are those in immunology ("blood" differences), in the study of disease (resistance and susceptibility), and in the investigations of growth and development. Although the first important study on variations in the effects of drugs (see Section III,D,2) was reported for rabbits, pharmacogenetic investigations are scarce. This fact is particularly regrettable because the genetics of the rabbit has advanced tremendously in recent years (for a review see Sawin, 1955). These developments are summarized briefly.

1. Genetic Studies in the Rabbit

Intensive efforts to develop inbred races of rabbits have been in progress for a considerable time; however the possibility of development of truly isogenic races is still uncertain, probably due to the accumulation of a large number of lethal genes resulting from the long period of outbreeding of domestic breeds.

Inbreeding of rabbits has been carried out for many years at the Jackson Laboratory to develop inbred strains and mutant stocks of biological value, and to study the effects of inbreeding. During this period there have been stillbirths, early mortality, various types of malformations, and a varied degree of infertility in some of the stocks. In a comparison of two stocks, in the BRY, which has been inbred for more generations than the CA, the incidence of deformities in the head, limb and sternum was greater than in the CA. The differences were interpreted as resulting from the segregation of genetic determinants or from their level of inbreeding, or a combination (Chai and Degenhardt, 1962).

The prospects of obtaining inbred races depends on (1) synthesis of stocks containing a maximum number of genes essential for high reproduction and a minimum of deleterious ones before inbreeding is initiated, and (2) selective inbreeding to rapidly fix those genes in the homozygous state. Efforts have indeed been made to breed rabbits so as to combine certain genes in the homozygous state, while others are still segregating from heterozygous carriers; in this way genetic tags are provided for use in a variety of studies and for purposes of economy in linkage tests. As of 1955, at least 32 specific loci, having one or more alleles, have been recognized. A tabulation of the mutated genes of the rabbit and their associations in the chromosomes has been made by Sawin and is reproduced as Table VII. It is obvious even from the brief phenotypic description that certain of these genes would lend themselves for pharmacologic investigations, although most of these lack the basic biochemical groundwork. Although polygenic variations are well known to influence a variety of systems, in the following sections certain selected single gene variations will be considered because of their greater potentialities for pharmacologic research; however, certain differences in behavior patterns (e.g., maternal and sex behavior) would afford unique material for genetic studies of hormone balance and hormone therapy.

D. Responses Controlled by Heredity in the Rabbit

TABLE VII
KNOWN MUTATED GENES OF THE RABBIT AND THEIR ASSOCIATIONS IN THE CHROMOSOMES[a]

Linkage group	Alleles	Phenotype
IV	A, a^t, a	Gray, black and tan, black
	Ac, ac	Normal lethal achondroplasia[b]
V	An, an	A antigen vs. α-agglutinin
VI	As, as	Atropinesterase present in blood serum, no enzyme
I ?	Ax, ax	Normal, sublethal ataxia[b]
I	B, b	Black, brown (chocolate)
V	Br, br	Normal feet, brachydactylia
	Bu, bu	Normal buphthalmos, or hydrophthalmos
I	C, c^{ch3}, c^{ch2}, c^{ch1}, c^h, c	Fully colored, chinchilla^3, chinchilla^2, chinchilla^1, Himalayan, complete albino
	D, d	Intense, dilute
	Da, da	Dachs-viable achondroplasia, normal
II	Du, du^d, du^w	Unspotted, dark Dutch, white Dutch
IV	Dw, dw	Dwarf,[b] normal size
VI	E^d, E, e^j, e	Steel (with A), gray (with A), Japanese, yellow
II	En, en	English (spotted), unspotted
VII ?	Ep, ep	Normal, epilepsy or audiogenic seizures
V	F, f	Normal, furless
	H^1, H^2, H^0	Hemagglutinogen 1 in blood
		Hemagglutinogen 2 in blood
		No hemagglutinogen in blood
II	L, l	Hair length normal, long-haired angora
	Lx, lx	Normal luxate
	Mc, mc	Normal, lethal muscle contracture[b]
	N, n	Normal naked
	O, o	Normal, osteopetrosis[b]
	P, p	Pelger,[b] normal
III	R^1, r^1	Hair length normal, short-haired (rex^1)
III	R^2, r^2	Hair length normal, short-haired (rex^2)
	R^3, r^3	Hair length normal, short-haired (rex^3)
	S, s	Normal coat, satin coat
	Sp, sp	Normal, spastic paralysis[b]
	Sy, sy	Normal, syringomyelia[b]
	T, t	Normal, shaking palsy (tremor)[b]
	V, v	Self-colored, self-white (eyes blue)
IV	W, w	Normal agouti (normal band), wide-banded agouti
	Wa, wa	Normal, hair waved
	Wh, wh	Wirehair, normal
	Wu, wu	Normal hair, wuzzy
I	Y, y	White fat, yellow fat

[a] From Sawin (1955).
[b] Lethal.

2. Genetic and Pharmacologic Properties of Atropinesterase in Rabbits

The first study showing a clear-cut difference in drug action owing to a genetically determined enzyme in the rabbit was reported by Glick and Glaubach (1941) and Sawin and Glick (1943). Previously Fleischmann (1910, quoted by Sawin and Glick, 1943) reported that certain rabbits could destroy atropine while others could not, and Levy and Michael (1938) suggested the possibility that atropinesterase is an inherited factor.

Sawin and Glick (1943) presented evidence that rabbits with blood serum and enzyme capable of hydrolyzing atropine inherit this peculiarity in a gene (As) borne in the same chromosome as the gene (E) for the extension of black pigment in the coat. The gene, As, is incompletely dominant, homozygotes producing the enzyme more effectively than heterozygotes. The enzyme is not present at birth, but appears at about one month of age, and tends to occur in greater concentration in females and to be demonstrable in a higher percentage of them than in males.

Atropinesterase splits l-hyoscyamine into tropic acid and atropine. The rabbit enzyme hydrolyzes, however, other tropine esters as well, e.g., homatropine and scopolamine (Glick and Glaubach, 1941); also it is able to split certain esters of morphine, e.g., monoacetylmorphine (Sawin and Glick, 1943) and is inhibited by high concentrations of physostigmine only. Therefore, atropinesterase is not specific in its ability to find atropine. The enzyme occurs throughout the body with highest concentrations occurring in the liver and intestinal mucosa; lack of activity was found only in brain and in aqueous humor of the eye. Among various breeds the enzyme was present in chinchilla rabbits and stocks of more mixed origin, but it was absent in New Zealand whites and in the A-race of Castle. In those breeds in which it was present, atropinesterase shortened greatly the pharmacologic response to the various compounds listed above.

3. Other Breed Differences in Drug Responses

A situation similar to that with atropinesterase exists in rabbits relative to the existence or lack of a *cocaine-esterase* which differs from

D. Responses Controlled by Heredity in the Rabbit

both cholinesterase and atropinesterase; aside from this biochemical observation, neither pharmacologic nor genetic studies have as yet been published (Ammon and Werz, 1959).

An interesting breed difference in the lactogenic response of the rabbit to *reserpine* has been reported (Tindal, 1960). Lactogenic effect of reserpine was described by Kehl *et al.* (1957); however, while it was confirmed in pseudopregnant and estradiol-primed New Zealand white rabbits (Meites, 1957), negative results were obtained in pseudopregnant Dutch rabbits (Benson *et al.*, 1958). The possibility of a breed difference was suggested: a single intravenous injection of reserpine (1 mg/kg) given either on day 15 of pseudopregnancy or after priming with 0.2 mg 17β-estradiol daily for 10 days, resulted in lactogenesis in virgin female rabbits of the New Zealand white breed. Thymus weight dropped significantly after reserpine injections in pseudopregnant rabbits, but reserpine did not lower the thymus weight in estradiol-primed rabbits to an extent greater than could be accounted for by the effect of the estrogen alone. In Dutch rabbits treated identically lactogenesis did not occur (Tindal, 1960). From these results it was concluded that there is a breed difference in *lactogenic response* of New Zealand white and Dutch rabbits to reserpine.

4. Hereditary Chondrodystrophy

A number of genes in different species induce abnormal or disproportionate growth; they display a similar sort of generalized pattern and primarily seem to differ only in the relatively greater expression of certain different localized portions of that pattern, indicating that there is some common element in the processes necessary to their fulfillment. The genes, when homozygous, cause abnormalities that are lethal at an early age in the Creeper fowl (Landauer, 1932), achondroplastic (gene symbol *ac*) rabbit (Pearce and Brown, 1945), and certain of the cattle dwarfs (Julian *et al.,* 1957); when heterozygous, they vary considerably in their expression from almost no to obvious abnormality, whereas animals such as the basset hound and dachshund (considered disproportionate dwarfs; Stockard, 1941) are able to live normal lives although homozygous. Between these two groups is the dachs rabbit (Sawin and Crary, 1957); in some genomes, the gene (symbol *Da*) may induce severe crippling, in others it permits survival to adulthood and production of viable young (although few of the young can be raised without a foster mother). Morphogenetic studies of the dachs or chondrodystrophic rabbit have been made by Sawin *et al.* (1959a, b). Together with the special growth characteristics induced by the recessive lethal achondroplasia (*ac*), primordial dwarfism (*Dw,* nanosoma vera), and the specific racial differences in growth pattern now available in the rabbit (Sawin and Crary, 1956), there is not only an unusual opportunity for analysis of possible genetic agencies that bring about changes in relative size, but also there is great opportunity for biochemical research.

The problems of hereditary chondrodysplasia in man have been considered in a recent book by Hobaek (1961). Indications are that hereditary chondrodysplasias occur so commonly as to be considered the most frequent and practically the most important group of systemic diseases of the skeleton (Marquardt, 1949); they include gargoylism, chondrodrystrophia fetalis (atypical chondrodystrophia, chondrohypoplasia), Spaet-Hurler syndrome, etc. Results of biochemical research in the mucopolysaccharide field in the last few years have opened new approaches to at least the beginning of an understanding of the etiology of these diseases. Even though a final solution of the etiologic problems still seems to lie beyond the immediate future and no definite conclusions can be drawn as yet, it seems probable that rapid advances in mucopolysaccharide (normal and pathological) biochemistry will be made, thereby preparing the ground for explaining the hereditary metabolic anomalies concerned.

5. Recessive Buphthalmos

Buphthalmos (hydrophthalmos; congenital infantile glaucoma) in rabbits has been known to occur sporadically for a long time; in fact the earliest report dates back to 1886 (Schloesser, 1886; for a review of all the literature see Hanna *et al.,* 1962). Buphthalmos has been of interest to European geneticists but has attracted little attention in the United States. Yet, because of its similarity to congenital glaucoma in humans, this condition is of particular interest to the field of experimental ophthalmology (L. Allen *et al.,* 1955; Burian *et al.,* 1960).

Buphthalmic rabbits have occurred since 1946 in the New Zealand white stocks maintained at the Roscoe B. Jackson Memorial Laboratory.

D. Responses Controlled by Heredity in the Rabbit

The occurrence of buphthalmos in litters appeared to be sporadic in repeated matings of the same carrier parents. Breeding and litter records showed that, in accordance with Geri (1954, 1955), buphthalmos in rabbits is a semilethal condition. A number of variables were found to affect the penetrance of the disorder; there was a deficiency of both buphthalmic and normal males in litters from carrier parents in which buphthalmic offspring occurred, and an excess of stillborn females in litters in which no buphthalmos appeared. The birth weight of buphthalmic females was significantly greater than that of their normal female siblings but the birth weights of normal and buphthalmic males did not differ. It was suggested that buphthalmos is a frequently occurring symptom of a systemic developmental defect in rabbits (Hanna et al., 1962). Although the primary defect responsible for the development of buphthalmos is not known, various experiments suggested an abnormality of the outflow mechanism resulting in an inability to maintain normal fluid relationships within the eye. This view was further supported by preliminary histologic findings demonstrating an absence of Fontana's spaces, the iris pillars, and either total absence or a rudimentary development of the trabecular canals and intrascleral channels (Babel, 1944). In view of the implications of both carbonic anhydrase and Na-K ATPase maintaining proper water and ion gradients, ion transport and secretion, it is important to ascertain whether or not buphthalmos can be prevented by treatments with acetazolamide (carbonic anhydrase inhibitor) and/or digitalis alkaloids (specific inhibitors of Na-K ATPase).

6. Epilepsy or Audiogenic Seizures

Epileptic-like seizures (ep) have long been known to occur occasionally in the Vienna white breed of rabbits; the pathological condition is apparently restricted to white, homozygous (vv) individuals and has never been observed in colored individuals derived from Vienna white crosses (Castle, 1940). However at the Jackson Laboratory it has occurred in a white spotted race (resembling the Vv animal) in which the blue-eyed whites (ve) segregate (presumably the Vienna white). Whether ep is a mutation at the Vienna white locus and thus one of three alleles (of which V is normal, o causes absence of pigment, and ve is the blue-eyed white which is subject to seizures) or whether it is a by-product of this gene requiring a particular genetic background is as yet uncertain.

E. Hereditary Characteristics of Guinea Pigs

Although attempts have been successful to produce inbred strains of guinea pigs (Carworth Farms) with very different characteristics (e.g., skin sensitivities to various chemicals, toxins, and toxoids), the raising of guinea pigs under rigid genetic control is time consuming and costly. There still are small colonies of inbred lines, but most guinea pigs are raised under "ordinary conditions." Also, although a number of mutant genes in the guinea pig have been recognized, they concern mostly color (pigmentation) and will not be discussed.

In recent years guinea pigs have become of particular interest because of the fact that they are unable to synthesize L-ascorbic acid (similar to man and other primates) and because of their high excretion of corticosteroids.

1. Inability to Synthesize Vitamin C (Scurvy)

Guinea pigs are unable to convert L-gulonolactone to L-ascorbic acid, a step which is catalyzed in the rat by enzymes present in the liver. They thus require vitamin C in their diet to prevent scurvy. The rat, a typical species that is independent of a dietary source of the vitamin, synthesizes L-ascorbic acid from D-glucose as follows: D-glucose → D-glucoronolactone → L-gulonolactone → L-ascorbic acid.

2. Corticosteroids in the Urine of Normal and Scorbutic Guinea Pigs

The isolation of cortisol (hydrocortisone) from the urine of guinea pigs substantiated the finding also of a normally high excretion of formaldehydogenic corticoids in the urine and the suggestion that the major corticosteroid excreted was hydrocortisone (for review see Burnstein et al., 1955). Ascorbic acid had been implicated in the biosynthesis of corticosteroids, an observation that is supported by the increase of biologically active cortisol-like material in the urine of scorbutic animals; 6β-hydroxycortisol was excreted in a concentration of 80%

F. Genetically Controlled Differences in Catalase

and cortisol was excreted in a concentration of over 300% above that in normal controls. ACTH treatment in scorbutic guinea pigs did not significantly increase the concentration of urinary steroids over that present in untreated scorbutic animals. However, in normal guinea pigs, ACTH significantly increased the urinary excretion of 6β-hydroxycortisol, 2α-hydroxycortisol, and cortisol. Lysergic acid diethylamide (LSD) was without effect, and, in combination with ACTH, did not alter the values observed with ACTH alone (Nadel et al., 1957). It may be inferred that either LSD was ineffective in producing psychotomimetic responses in the dosage administered or that the guinea pig metabolizes LSD in a manner different from man and other animals.

F. Genetically Controlled Differences in Catalase Activity in Red Blood Cells of Cattle, Dogs, Guinea Pigs, and Man

Acatalasemia is a rare hereditary constitutional abnormality characterized by a deficiency of catalase in the blood. Genetically, hypocatalasemia appears to be heterozygous, or a carrier state of the acatalasemic gene; catalase levels are about one-half that of normal blood.

A similar condition occurs in guinea pigs (Radev, 1960) and certain breeds of dogs (Allison et al., 1957). In dogs, the disorder was discovered accidentally during the course of experiments in vitamin E deficiency; it was noticed that in the presence of *dialuric acid,* erythrocytes of some dogs became brown within a few minutes while those of others maintained their normal red color for $\frac{1}{2}$ hour or longer. It was supposed that in the presence of sufficient catalase, *hydrogen peroxide* (produced by autoxidation of dialuric acid) might accumulate and oxidize the hemoglobin to methemoglobin which is brown; however, if there were sufficient catalase in the erythrocytes, they would retain the red color of oxyhemoglobin. Measurements of catalase activity in erythrocytes revealed that dogs whose red blood cells had become brown in the presence of dialuric acid had catalase levels of less than 1.5 millimole/ml cells/min; in addition it was found that the mean erythrocyte catalase level of dogs is about 1/30 that in human erythrocytes. Studies on the inheritance of catalase levels consisted of determining catalase in offspring of various matings: high × high, high × low, and low × low. Results were consistent with a single mode of inheritance; it was supposed that there is a single pair of alleles, C and c, and that in the presence of C more catalase is produced than in the presence of c. Based on this hypothesis three genotypes would be expected, CC, Cc, and cc; in fact, dogs could be classed according to this scheme having levels of above 2.5 millimole/ml/min (CC), below 2.5 millimole/ml/min (Cc), and less than 1.5 millimole/ml/min (cc).

Similar evidence for a single pair of genes controlling catalase level was reported for cattle and guinea pigs (Putilin, 1929); again the activity of the heterozygotes is intermediate to those of the two homozygotes.

It is noteworthy that catalase deficiency in dogs may be confined to the erythrocytes: tests on liver homogenates from dogs with low erythrocyte-catalase levels showed the same activity as liver homogenates from dogs with high red blood cell catalase. Also, in dogs with low catalase no signs of hemolysis, liability to infection, or other disability are observed; in contrast, acatalasemic man is subject to progressive oral gangrene.

Recently a series of experiments on the "catalase-protein" of acatalasemic red blood cells of man have been reported (Takahara et al., 1962). Three possibilities were considered to obtain information on the precise nature of the biochemical defect: (1) catalase may be totally absent, (2) it may be present but inhibited, and (3) a substitute catalase-like molecule may be enzymatically inactive because of a structural defect in the molecular arrangement of the enzyme. Since the data presented indicate that an extract of acatalasemic blood had almost no reactivity to catalase antibody, it could be assumed either that the protein does not exist or, if present, it is so highly denatured as to possess no common determinant group. Further studies are underway (Takahara et al., 1962).

G. Excretion of Uric Acid and Amino Acids in the Dog

In the following subsections, two spontaneous conditions involving abnormal urinary metabolites in dogs will be described briefly. The genetic aspects of high urinary acid excretion are well known, while a genetic basis for cystinuria is very suggestive from the breed and sex distribution.

G. Excretion of Uric Acid and Amino Acid

1. High Uric Acid Excretion

The metabolism of purines is known to yield different end products in different species of animals. In some (man, chimpanzee, and birds), uric acid is the substance excreted in largest quantities; others (certain monkeys, various ungulates, marsupials, carnivores, and rodents) excrete allantoin as the principal end product of purine metabolism (Hunter and Givens, 1914, quoted by Benedict, 1916). However, Benedict (1916) reported an important exception in this classification when he found a Dalmatian dog who excreted in its urine as much uric acid as an adult man. Wells (1918), in studies of the enzymes of purine metabolism, detected the presence of uricase in the liver of a Dalmatian dog. Since, in this respect, the Dalmatian did not differ from other breeds of dogs, this unique capacity for excreting uric acid could not be explained by lack of purine oxidation to allantoin.

A number of investigations concerned with the heritability of high uric acid excretion in Dalmatian dogs (Onslow, 1923, quoted by Trimble and Keeler, 1938) showed that it is inherited as a recessive, autosomal character (Trimble and Keeler, 1938). "High producers" of uric acid are dogs that on a purine-free diet excreted 28 mg per day or more of uric acid. In addition it was found that "high uric acid excretion" is not associated with the gene complex producing the "harlequin" or Dalmatian type of spotting (pigmented patches or spots).

Renal uric acid calculi are quite frequent in the Dalmatian dog (Meier, personal observation); this breed provides excellent material for pharmacologic investigations relative to purine metabolism and renal function, calculus formation, prevention and therapy.

2. Amino-Aciduria in Canine Cystine-Stone Disease

The urine of dogs with cystine-stone disease shows some of the characteristics of human cystinuric urine.* These include a grossly ab-

---

* Genetical analysis of cystinuria (Wollaston, 1810, quoted by Harris, 1959) revealed a rather complex situation. On genetical grounds two main forms of classic cystinuria are differentiated: the "recessive" and the "incompletely recessive"; whether or not there are several genes causing "recessive cystinuria," one of them resulting in the "incompletely recessive" condition, or different alleles remains to be discovered (Harris, 1959).

It should be remembered that cystine is excreted in unusual amounts in a number of other genetically determined conditions, i.e., cystinosis and the different variants of

normal amount of lysine and a very substantially elevated cystine level; arginine, and in some cases ornithine, is present in smaller, but still abnormally large amounts. The plasma-amino acid pattern is normal in respect to these basic amino acids and cystine (Treacher, 1962). The presence, also, of citrulline in the cystine-stone disease urines raises the possibility that citrulline, being structurally similar to arginine, and somewhere between cystine and ornithine in basicity, is normally reabsorbed from the kidney tubule by the same mechanism or a part as is involved in the reabsorption of the basic amino acids and cystine. The amino-aciduria in the dog is thus of the "renal" and not the "overflow" type.

The heritability of cystine-stone disease in dogs has not been studied; however, from personal observations it is known to occur in dachshunds. In an analysis of 220 cases of urinary calculi in dogs, 34 cases or over 15% consisted mainly of cystine. All of the cases were males (Treacher, 1962).

H. Idiopathic Familial Osteoporosis in Dogs and Cats: "Osteogenesis Imperfecta"

Osteogenesis imperfecta is a disease that occurs in small animals, especially cats, as well as in man (Schnelle, 1950). In cats, it is most prevalent among Siamese, and in dogs it has been observed most frequently in poodles and Norwegian elkhounds. Although no genetic investigations have ever been made, a hereditary basis is indicated by the breed distribution and the increased occurrence in certain litters and families. Diagnosis of the condition rests on the marked thinness of the bone cortices, pathological fractures, rapid healing with little callus formation, striking flaccidity of the ligaments and joints, and the spontaneous improvement at puberty.

It may be pharmacologically significant that in *dogs* osteogenesis imperfecta closely resembles copper deficiency. Although from the experimental procedures employed in studies on osteogenesis imperfecta, it seems unlikely that it is pathogenetically identical with copper deficiency (despite a slightly reduced serum copper level; liver copper

---

the Debré-de Toni-Fanconi syndrome, and Wilson's disease or hepatolenticular degeneration. These may be differentiated on clinical, biochemical, and genetical grounds (more generalized amino-aciduria, rarely cystine-stone disease).

levels were normal); however it is likely that some alteration in copper metabolism is the cause of the defect in osteogenesis imperfecta although perhaps not directly related to that of copper deficiency (Calkins et al., 1956).

In *cats,* experimental reproduction of osteogenesis imperfecta is possible by feeding a low calcium diet consisting of either beef or sheep heart (Ca:P=1:20). The ensuing negative Ca-balance is reversed by calcium gluconate, calcium borogluconate, or calcium carbonate as well as iodine (as low as 100 μg/day) even without added calcium. From this latter evidence a thyroid interplay in the pathogenesis of osteogenesis imperfecta was inferred; indeed, on a low calcium diet there is histologic evidence of thyroid hyperactivity, e.g., little or no colloid, small or no follicular structures, and high columnar epithelium (Scott, 1961, personal communication). The involvement of the thyroid may also be inferred from the fact that osteogenesis imperfecta in man occurs often in association with thyrotoxicosis.

I. Amine Content in Adrenal Glands of Families of Cats, and General Aspects of Catecholamine Storage in Animals

Butterworth and Mann (1957) have obtained a range of 13 to 91% of noradrenaline in the adrenal glands of cats. This animal-to-animal variation is in marked contrast to that in most other laboratory animals where the proportion of noradrenaline to adrenaline is constant for a given species. While there is wide variation between individual animals in both sexes, there is no significant difference between the percentages of noradrenaline in male and female cats. Also there is a close relationship between the total amine content of the two glands in any one cat when expressed as per gland then when expressed as per 100 mg of gland. No difference exists between the left and right adrenal gland in the percentage of noradrenaline (Butterworth and Mann, 1957). Although there is a wide range in the percentage of noradrenaline (8–84%) between litters of cats, there is little or no variation within a litter independent of litter size and sex distribution. Also, as might be expected from the mother-litter relationship, the values for successive litters from each mother are similar (Butterworth and Mann, 1960a, b). The application of these results is that it is possible to perform individual experiments on littermate cats; also one gland of each animal may serve as the control for the other.

From a variety of observations two independent types of cells are responsible for the secretion of the two adrenal medullary hormones. However, their cytological properties are subject to considerable species-dependent differences. For example, the histochemical reaction for acid phosphatase is negative in the noradrenaline-containing cell islets of the rat, but strongly positive in the adrenaline-containing cells; the same technique fails to differentiate between the two cell types in the adrenal medulla of the mouse, cat, dog, and cow. Also, there is species variation in the pattern of the noradrenaline-containing cells. In most animals these cells are clearly delineated islets within the main bulk of the medulla, but in the hamster they are exclusively peripheral and in the mouse their distribution is random; the guinea pig and rabbit lack specific cell groups. However, in spite of differences between species, the distribution pattern is fairly constant within a given species. Pharmacologic evidence for the existence of two cell types consist of the following findings: insulin completely depletes the medulla of adrenaline, renders the chromaffin reaction negative safe for positive islets, and affects neither the noradrenaline content nor the histochemical reactions in the noradrenaline-containing islets of the rat. On the other hand, reserpine in a low dosage causes a selective loss of noradrenaline and renders negative the chromaffin, fluorescence, and iodate reactions in the noradrenaline-containing cells of the rat. Selective hyperplasia and high cholinesterase activity of the noradrenaline-containing cells is produced in the rat by prolonged administration of nicotine; in the mouse, thiouracil causes an increase in size and hormone content of the noradrenaline-containing cells (for a review of these findings see Eränkö, 1961).

The problem of how the large amounts of catecholamines are found within granules of the adrenal medulla is intriguing since their quantities equal those of proteins and lipids. The general chemical composition of high-density granules shows that amines, adenosine phosphates, proteins, lipids, and water constitute about 98% of the granule wet weight. It is now well established that the amines are stored with an equivalent amount of adenosine phosphate, mainly ATP. The content of adenosine phosphates shows considerable species differences (Graph IV). ATP may play a dual role: it may be structural in part and, in view of the high granular content of ATPase activity, it may also be involved in the release of the amines (for a review see Hillarp, 1961).

Few data seem to be available concerning the normal content of adrenaline and noradrenaline in organs of mice. Owing to the fundamental differences in the techniques for measuring them, comparison and evaluation of the results are extremely difficult. However, one investigation relating to the adrenaline and noradrenaline content of

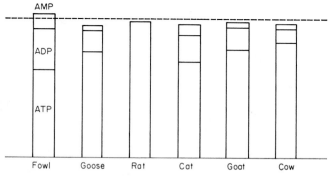

GRAPH IV. Granular content of adenosine phosphates. The ratio of the sum of equivalents of adenosine phosphates to equivalents of catecholamines is shown by the height of the columns. The dotted line equals a ratio of 1. (From Hillarp, 1961.)

adrenal glands, heart, liver, and spleen of normal adult white inbred mice is worth mentioning; the amounts of compounds have been measured fluorometrically, and also the influence of various pharmacological substances have been determined (DeSchaepdryver and Preziosi, 1959). Of interest is the finding that in some extracts of heart and spleen no adrenaline could be detected, a fact suggesting that adrenaline may not be considered a regular constituent of these organs in mice and that its occurrence may depend on the presence of chromaffine cell groups which are irregularly scattered throughout the organism. It is known that administration of insulin, nicotine, histamine, and reserpine may provoke a lowering of the catecholamine content of the mammalian adrenal gland; with regard to their catecholamine-depleting properties in adrenal glands, heart, liver, and spleen in mice these substances may be graded in order of decreasing activity as follows: reserpine, insulin, nicotine, and histamine. It is noteworthy that after depletion of adrenaline and noradrenaline stores, the time required for restoration of the catecholamines may be remarkably long and that the rate of resynthesis apparently depends on the nature of the substance used and not on the degree of depletion. Of the two reserpine-like drugs used in this study only the one with sedative action, i.e., desmethoxyreserpine, provoked a complete depletion of adrenal gland noradrenaline, while the amount of this catecholamine was only transiently lowered after a dose of reserpiline, which is supposedly devoid of tranquilizing properties. However, it must be noted that sedating doses of chlorpromazine, mepazine, perphenazine, and promazine provoked only a slight depression of the adrenal gland content, as did sedating doses of meprobamate. This then would imply that sedation may presumably occur without depletion of noradrenaline, at least in the adrenal glands, and possibly also in other tissues, since among the organs with adrenergic innervation the adrenal glands appear to be most sensitive in this respect. Furthermore, the experimental data presented render distinct proof for the possibility of a preferential depletion of adrenaline and noradrenaline after pharmacological stimulation. A striking feature of the findings reported relates to the fact that blockade of the monamine oxidases by iproniazid which is known to largely prevent catecholamine depletion after reserpine, also unequivocally exerted a protective effect against adrenal gland catechol depletion provoked by other pharmacologic agents. According to the pharmacologic substances used one or the other of two mechanisms may prevail, i.e., "secretion" and "liberation with subsequent inactivation" of catecholamines; definite predominance of the "liberation-inactivation" mechanism occurs in the case of reserpine.

J. Inherited Traits in Farm Animals (Cattle, Horses, Swine, Sheep, Chickens) and Dogs of Potential Pharmacologic Interest

Through the years a fairly impressive number of inherited traits (lethal, semilethal, deleterious, or undesirable) have been reported in farm animals. A list of considerable length has been prepared by

Stormont (1958). While they are of great economic importance, few may be of interest from a pharmacologic viewpoint. It is surprising, however, that, in spite of extensive use of dogs in experimental biology and medicine, not more genetic studies have been made of numerous anomalies that may have considerable significance in bio-medical research (Burns, 1952; Schuman, 1955).

A few selected traits in farm animals and dogs are presented in detail throughout this text; certain others that pharmacologists may have occasion to investigate are briefly listed below.

Cattle

> Congenital porphyrinuria, a semilethal trait, commonly called "pink tooth." Excess of coproporphyrin and uroporphyrin; animals photosensitive and developing lesions in unpigmented areas of skin. Shorthorns and Holstein-Friesians (Fourie, *OJVSAI** 13:383, 1939; Jorgenson and With, *N* 176:156, 1955; Clare, *AVS* 2:191, 1955).
> Night blindness (Craft, *JH* 18:215, 1927).
> Dwarfism of various types (Mead *et al.*, *JH* 33:411, 1942; Gregory *et al.*, *Hi* 22:417, 1953; Johansson, *H* 39:75, 1953).

Horses

> Congenital blindness (Schnelle, *DTW* 58:325, 1951).

Swine

> Congenital porphyria (dominant type). Affected animals characterized by reddish-brown color of bones and teeth (Jorgenson and With, *N* 176:156, 1955).

Sheep

> Dwarf. Thyroid disturbance leading to death within a month after birth (Bogart and Dyser, *JAS* 1:87, 1942). Congenital photosensitivity. Affected animals thrive with feeding on grass;

---

* Literature Abbreviations: *OJVSAI*, Onderstepoort Journal of Veterinary Research; *N*, Nature; *AVS*, Advances in Veterinary Science; *JH*, Journal of Heredity; *H*, Hereditas; *Hi*, Hilgardia; *BKAES*, Bulletin (533) Kentucky Agriculture Experiment Station; *DTW*, Deutsche tierärztliche Wochenschrift; *JAS*, Journal of Animal Science; *NZJSTA*, New Zealand Journal of Science and Technology; *G*, Genetics; *VR*, Veterinary Record.

ears, eyelids, and lips become edematous and bleeding. Death inevitable except when protected from sunlight. Condition similar to facial eczema in that surface lesions depend on presence of phylloerythrine; however, liver not damaged. (Southdowns) (Hancock, *NZJSTA* **32**:16, 1950).

Chickens

Jittery. Sex-linked lethal with retraction and rapid shaking of head in newly hatched chicks (Bohren, *G* **35**:655, 1950).

Shaker. A sex-linked semilethal character similar but less extreme than jittery (Scott *et al.*, *JH* **41**:254, 1950).

Dogs

Retinal atrophy. Irish setters, Gordon setters, and Labrador retrievers. Atrophy of retinal cells resulting in night blindness at 8 weeks of age (Hodgman *et al.*, *VR* **61**:185, 1949; Young, *VR* **67**:15, 1955). Hemophilia (see below).

1. Non-homology of Cattle and Mouse Dwarfism

Numerous distinct dwarf mutants of cattle and mice have been reported. The comparative genetic and biological nature of the mutants in the two species are of considerable interest and biological importance, including the problem of genetic homology.

The general morphologic characteristics of the bovine mutant have been described (Johnson *et al.*, 1950; Gregory *et al.*, 1951); it is subnormal in size and belongs to an achondroplastic complex (Julian *et al.*, 1957; Tyler *et al.*, 1957). All components of the complex (brachycephalic dwarf) exhibit lesions of achondroplasia in the axial and/or appendicular skeleton (Gregory *et al.*, 1960a, b). In the brachycephalic dwarf, the pituitary-thyroid axis is reported to be normal (Crenshaw *et al.*, 1957), growth, gonadotropic and ACTH activities in the pituitary tissue either being normal or nearly normal (Marlowe and Chambers, 1954; Marlowe, 1960); sensitivity to insulin is greater than in normal cattle (Foley *et al.*, 1960). There is disagreement, however, as to whether the pituitary tissue is deficient in thyrotropic hormone activity (Caroll *et al.*, 1951); while body weight increases with age, the skeletal growth does not, even in response to hormone

J. Inherited Traits in Farm Animals and Dogs

therapy such as thyroprotein, testosterone, and stilbestrol or their combination.

The *mouse* dwarf mutant, A *dw* (Fig. 11), suffers from a thyroid-pituitary deficiency (Snell, 1929, Mollenback, 1940; Grunnet, 1942; Francis, 1944, 1945): although the pituitary has high concentrations of gonadotropic hormones, its growth hormone activity is difficult to

FIG. 11. Dwarf mouse, *dwdw* (forefront), with normal littermate. (Courtesy of Photo- and Art Department, Roscoe B. Jackson Memorial Laboratory.)

detect (Smith and MacDowell, 1930, 1931). It is similar to the bovine dwarf in being quite sensitive to insulin (Mirand and Osborne, 1953).

Thus from the available information it is certain that the Snell dwarf mouse is impaired in its pituitary-thyroid relationship and the bovine dwarf belongs to an achondroplastic complex with a normal thyroid-pituitary axis; this in itself would indicate that the pituitary-deficient mouse mutant is neither homologous nor isogenic with the bovine mutant. Since dwarf mice receiving injections of pituitary extracts from achondroplastic cattle stimulated growth as much as extracts from normal cattle supports the hypothesis that the two types of mutations are dissimilar (Carroll and Gregory, 1962).

Pertinent to this context are some studies in human pituitary dwarfs relative to their serum growth hormone content (Girard et al., 1961). Immunologic assays (by means of a human growth hormone rabbit antiserum) revealed that hypopituitary dwarfs had concentrations well below the lowest values encountered in normal children and adults (6–10 μg/100 ml). Values in normal adults (26.0 ± 5.53 μg/100 ml) are significantly lower than the average during childhood (39.6 ± 10.48 μg/100 ml). No accurate measurements have been made in pituitary dwarf mice and rats (Evans and Simpson, 1951).

2. Heredity and Variation in Domestic Fowl

The study of many mutations has shed much light on the gene complex of domestic fowl. An entire book devoted to the "Genetics of the Fowl" (Hutt, 1949) considers what characters are inherited, how they are inherited, how they affect the birds, how to eliminate them if undesirable, etc.; most importantly discusses the genetic resistance to disease and the genetic differences in nutritional requirements; and also lists all known mutant genes in addition to giving linkage groups and a revised chromosome map.

It may suffice here to acknowledge the existence of different breeds of varieties of fowl, e.g., Leghorns, Minorcas, Rhode Island Reds. A breed is a group of fowls related by descent and breeds true for certain characteristics; within such a breed there may be variations in color, comb, or other characteristics, but there is no specific number of differences necessary to differentiate varieties or breeds. While inbreeding is practiced by some poultry breeders, the objectives may differ, but they relate especially to color varieties or they are of an economic nature. Although inbred strains of chickens, according to the definition applied to inbred mice, do not exist, the term "strain" is nonetheless used.

It seems reasonable to expect differences in pharmacologic responses between strains and breeds of chickens. This relates particularly to evaluations of drugs that are tested as coccidiostats and others. Two parameters appear to be particularly relevant: (1) therapeutically effective dosage levels of a drug or its toxicity and (2) problems relating to residues of compounds (similar to food additives) in edible tissues of poultry (e.g., amprolium or 1-[(4-amino-2-n-propyl-5-pyrimidinyl)methyl]-2-picolinium chloride hydrochloride, sulfaquinoxaline, nitrophenide, nitrofurazone, nicarbazin, and glycarbylamide). While these aspects will not be dealt with, it may be of interest, as an illustrative example, to refer to differences in copper retention in two strains of chickens (Lillis et al., 1963). The chickens used were two strains of white Leghorns, one (designated R line) had genetic resistance to the avian leukosis complex, the other (S line) had susceptibility. Cupric acetate was injected intramuscularly at 0.91 and 9.1 mg/kg, respectively. While no differences were observed at the lower dosage, significantly more copper was retained in the liver by the R line than S line at the higher level. Also, at 12.5 and 15 mg/kg, respectively, the mortality was significantly greater in the S line, the chickens displaying characteristic symptoms of copper poisoning described for guinea pigs by Huber (1918). Analogous to the findings with copper, injection with phenylmercuric acetate or mercury chloride, revealed greater mercury retention by R line livers than S line livers (Miller et al., 1959); these data suggest a similarity in the detoxification of copper and mercury following intramuscular injection, but do not bear upon the relationship to the relative susceptibility to avian leukosis.

K. Species Dependence of Induced Lesions

Although the mechanism of action of drugs in experimental animals is assumed to be analogous to that in man, it may or may not be identical. It is obvious now that any information forthcoming from experimental approaches must be interpreted first in terms of the physiology of the host animal(s). In view of the existing species and strain specificities of various reactions, the test systems in animal experimentation should be carefully scrutinized. The following selective examples may be illustrative.

One of the most provocative problems in atherosclerosis research is the cause of species differences in susceptibility to this disease. Spontaneous atherosclerosis is rare in animals but has been reported to occur in certain strains of pigeons and the baboon. Since a large body of evidence implicates the lipoprotein complex in this disease, differences in lipoprotein binding relative to quantity in various animals may be an index of species susceptibility. In determinations of the "readily extractable cholesterol" (REC) as a parameter of evaluating the stability of the lipoprotein complex, the following results were obtained; the REC in the dog is 5.8% of the total cholesterol, the rabbit is 2.9%, the rat 2.9%. The baboon, in spite of blood cholesterol similar to the dog, has an REC of 13.9%, nearly 2½ times as much, while man under similar conditions has a value of 49.0%. These results suggest a relationship between the stability of the lipoprotein complex as manifested by the REC and the frequency of atherosclerosis in different species. The REC is independent of an increase in the absolute β-lipoprotein fraction (Sherber, 1962).

It is well established that atherosclerosis can be induced readily in rabbits by the administration of a cholesterol-containing diet, but that diets containing cholesterol alone are without deleterious effects under similar conditions in the rat (Katz and Stamler, 1953; Deuel, 1955). While Schlichter and Harris (1949) suggested that the difference in response might be related to variations in the two species in the adequacy of the vasa vasorum of large arteries or to differences in lipid and cholesterol metabolism (Cook and Thomson, 1951a, b). More recently, variations in thyroid function have been implicated (see above): in the rabbit, treatment with a reticulo-endothelial-blocking agent, e.g., thorotrast or carbon, resulted in thyroid hypoplasia (Nathaniel et al., 1963), and in the rat similar treatment caused thyroid hyperplasia (Frankel et al., 1962). A similar effect is also obtained in the latter species by feeding cholesterol at a 1% level in the diet (Bernick and Patek, 1961). Thus there appears to be a correlation between the occurrence of atherosclerosis and the type of thyroid gland response; also it seems likely that a relationship may exist within any single species between the functional-morphologic character of its thyroid gland, the response to cholesterol feeding, and the susceptibility to atherosclerosis. It is, therefore, worthwhile considering the species dependence of thyroid function.

Studies on the 24-hour I^{131}-uptake, radiochromatographic distribution, and histologic appearance of the thyroids of BUB mouse fetuses of different ages revealed the following findings: development of thyroid function takes place between 15 and 17 days of age; accumulation

K. Species Dependence of Induced Lesions

of iodide from the circulation, organic binding of iodine, and formation of colloid appear between 15 and 16 days, whereas the ability to produce thyroxine and the formation of follicles appear between 16 and 17 days. Therefore, colloid formation is primary to follicle formation and production of thyroxine begins later than the collection and organic binding of iodine (van Heyninger, 1961); this finding is similar to that in rats (Gorbman and Evans, 1943) and rabbits (Waterman and Gorbman, 1956).

Since in the adult thyroid the organic binding of iodine takes place in the colloid (Wollman and Wodinsky, 1955), it might be expected that in the fetal thyroid, colloid would be present when the first organic binding of iodine occurs. This holds true for the fetal calf in which thyroxine is present a number of days prior to the formation of colloid droplets and follicles (Koneff et al., 1949). A similar situation is met for pigs (Rankin, 1941). Among adult animals, there is considerable variation in the morphologic appearances of the thyroid gland (Bernick et al., 1962). The normal gland in the rabbit and guinea pig contains follicles lined by low cuboidal epithelium, and intracellular colloidal droplets are not demonstrable by the periodic acid-Schiff (PAS) technique. In contrast, the thyroid follicles in the rat and the Golden hamster are lined by high cuboidal cells containing fine cytoplasmic PAS-positive droplets; I^{131} incorporation into thyroglobulin is exceedingly rapid in the latter species (more than 50% of the total thyroid I^{131} in 0.5 to 2 minutes) and is very slow in the former (50% value reached after 30 minutes) indicating greater activity in the rat and hamster than the guinea pig and rabbit (Pitt-Rivers, 1960). Cholesterol feeding induced thyroid hyperplasia in both rat and hamster with the follicular epithelium becoming columnar and containing an increased number of PAS-positive droplets; the amount of stored colloid is reduced. A diametrically opposite response occurs in the rabbit and guinea pig; the follicles are lined with squamous epithelium free of colloid droplets, indicating a hypoplastic state. It may be concluded that rabbit and guinea pig have a relatively low thyroid activity; and their livers being unable to efficiently metabolize exogenous lipids or cholesterol, these animals, therefore, show both hyperlipemia and hypercholesterolemia, conditions which are factors in their susceptibility to atherosclerosis (Bernick et al., 1962).

A sizeable body of circumstantial evidence has now accumulated implicating elevated serum lipid levels (or any given component, e.g., cholesterol) as at least one contributing cause of atherosclerosis. Therefore, one of several feasible chemotherapeutic approaches relates to control of serum lipid levels; these approaches have been admirably summarized by Steinberg (1962). In this context, only the therapeutic use of thyroid hormones and analogs pertains; while the exact mechanism by which thyroid hormones exert their effects in reducing hyperlipemia is not known, their clinical effectiveness is unequivocal. However, from the foregoing discussion it is not necessary to elaborate on the fact that equi-therapeutic dosages must have varying effects in different experimental animals. Since in hypothyroid rats, excretion of bile acids is reduced it is presumed that the primary effect of thyroid hormones is to increase degradation and excretion out of proportion to the increase in the rate of synthesis (probably a secondary compensatory response); cholesterol is disposed of either by excretion as such or in the form of bile acids formed in the liver by oxidation of the side chain. Some are converted to steroid hormones, but this is quantitatively not an important pathway. Acceleration of bile-acid formation from cholesterol or acceleration of the rate of excretion of cholesterol itself or of bile acids could reduce the availability of cholesterol.

To reduce the hypermetabolic activity (basal metabolic rate, BMR) of thyroid hormone, analogs of thyroxine have been developed which only lead to depression of cholesterol levels; among these are L-triiodothyronine (T_3) and triiodothyroacetic acid (TRIAC). A similar (as well as dissimilar) action on compounds is exerted by a recently obtained agent, choletyramine (MK-135), which increases bile-acid excretion by preventing intestinal reabsorption (Tennent et al., 1960).

In recent years the assay of progestational activity utilizes induced alterations in uterine carbonic anhydrase. Compared with the uteri of other species, the uteri of mice show a rather high carbonic anhydrase activity, with values ranging from 25 to 110 enzyme units per gram of uterine tissue (EU/gm) under various experimental conditions.

In diestrus, proestrus, and metestrus the carbonic anhydrase activity of the mouse uterus was almost the same (about 60 EU/gm); during estrus it was significantly increased (98 EU/gm).

Eight days after castration, uterine carbonic anhydrase activity was

K. Species Dependence of Induced Lesions

not significantly altered (47 EU/gm). Daily, subcutaneous treatment of spayed mice with 0.25 mg progesterone (pregn-4-ene-3,20-dione) did not have any significant effect upon uterine carbonic anhydrase (49 EU/gm). Treatment with estradiol monobenzoate (0.025 and 0.1 µg daily, subcutaneously, for 7 days) significantly increased the carbonic anhydrase activity (to 109 and 86 EU/gm, respectively). This effect of estrogen was counteracted by simultaneous administration of progesterone.

These results differ markedly from the results obtained in similar experiments with rats and rabbits in which, during estrus and estrogen treatments of spays, carbonic anhydrase activity is lowered. Also while in mice progesterone does not increase enzyme activity and in fact counteracts the enhancing effect of estrogen, progestational properties of a given substance are determined in rabbits from the elevation of carbonic anhydrase. Unexpected in mice is the observation that traumatic deciduomata have significantly less carbonic anhydrase activity per gram of uterine tissue (26 EU) than the control horns without deciduomata (48 EU). No explanation for the species difference may be offered at this time (Madjerek and van der Vries, 1961).

Of course, by no means is it to be inferred that induced pathological situations (e.g., triton-cholesterol hyperlipemia, alloxan-hyperglycemia, gold-thioglucose obesity, are not of the utmost importance in assessing potential therapeutic drug action and metabolism. A great deal has been learned thus far, and further exciting developments may be confidently anticipated, e.g., in the control of hyperlipemia and atherosclerosis the clinical (man) effectiveness of several drugs (e.g. nicotinic acid, heparin, and heparoids) is well established. Also, it should be remembered that aside from numerous dissimilarities between species, for certain characteristics there are a great many similarities in reaction. For example, enlargement of the salivary gland occurs in both mice and rats following treatment with large doses of N-isopropylnoradrenaline (isoproterenol) which is obviously a "general" effect of the catecholamine (Selye et al., 1961; Brown-Grant, 1961).

L. Potential Significance of Mutagenic Drugs to Man and Other Species

Mutations are changes in nucleotide sequence. The spontaneous rates of such changes have been estimated in different organisms with varying degrees of indirectness, e.g., the most frequently reported rate in man is between 10^{-4} and 10^{-5}. Whether or not so-called spontaneous mutations are actually induced has yet to be proved. Separation of the two types of mutations may, however, be justified in view of the fact that in spontaneous mutations the frequency of the primary event is not known exactly while with induced mutations, measurements can be made regarding the effect of a given treatment on the frequency of mutants.

Pharmacologic research is still the source of an ever-increasing number of chemical mutagens. Mustard gas was discovered because, pharmacologically, it resembles X-rays in the kind of burns it produces. Simply on pharmacologic grounds, other substances were found to be mutagenic; for instance, allyl isothiocyanate was tested because it was vesicant. Most carcinostatic agents are also mutagenic, e.g., the so-called alkylating agents, the connecting link between the two properties lying in the ability to break chromosomes. Also on pharmacologic grounds, a number of alkaloids have been tested successfully for mutagenic ability, e.g., scopolamine and morphine. Certain pyrrolizidine alkaloids (e.g., heliotrin), of concern to sheep breeders in Australia because they cause liver disease, when tested in *Drosophila* were found to produce very high mutation frequencies. A long list of various compounds that are used therapeutically could be added to those already mentioned as being mutagenic. It is clear, however, that, although there is as yet very little concrete information available, the exposure of man and other species to a chemical environment (drugs) may represent a definite genetic hazard. While in recent years overemphasis lay on the genetic hazards of radiation, there has been underemphasis on the possible genetic hazards of chemical mutagens. As has recently been discussed at a special symposium sponsored by the Josiah Macy, Jr., Foundation (Schull, 1962), drugs as well as common substances and natural metabolites such as formaldehyde, hydrogen peroxide, and caffeine may have potential significance as mutagens in man and other

L. Significance of Mutagenic Drugs

mammalian species. Caffeine (orally in the drinking water as a 0.3% solution) may be mutagenic in its effects on male fertility of mice, but was ineffective in producing translocations that cause semisterility (Cattanach, 1962, personal communication). In other tests for mutagenicity of caffeine, specific locus mutation rates in male and female mice (having 0.1% caffeine dissolved in their drinking water up to the age of 10 weeks) were compared. Their parents had the same treatment from the time of mating, so that the tested germ cells might be exposed to caffeine during embryonic development. The mutation rates did not differ significantly from each other, or from the known spontaneous rate; thus there was no evidence for induction of mutations by the caffeine treatment. Neither was there evidence for the induction of dominant lethals in the males. The treatment did not noticeably affect reproduction, but some mice developed aggressive tendencies toward their cage-mates. Although some mice were kept on 0.1% caffeine throughout their life, they continued to breed satisfactorily (Lyon *et al.*, 1962, personal communication).

Obviously, the genetic apparatus of mammals is not wholly unique, and extrapolation from a variety of other species (e.g., microorganisms) may be justifiable. Although there is no direct evidence for an effect of any chemical mutagen in man or mammalian organisms, except for a very few experiments with alkylating agents in mice, it would seem improper to dismiss chemical mutagenesis. In general, when cancer is treated with alkylating agents or other compounds which are known to be mutagenic, the genetic hazard can be discounted if the patient is beyond the reproductive age. However, there are instances, such as in Hodgkin's disease, where the individual is often within the reproductive age. This suggests an area for investigation as to whether patients (or animals) so treated, and who subsequently have offspring, will show genetic damage.

An interesting relationship exists between carcinostatic, mutagenic, and, curiously, carcinogenic agents; some substances have all three of these actions. Compounds may owe their anticancer and mutagenic effects to the same or similar action, for example, to interference with chromosome replication. At any rate, they do not uniquely attack cancer cells, but produce toxic (and even detrimental) changes in normal and usually rapidly growing tissues.

Another aspect of chemical mutagenesis relates to the question of teratogenesis, i.e., the induction of abnormalities in the developing fetus (see above). A number of teratogens are also known to be mutagens and it is conceivable, although entirely unproven as yet, that in some cases developmental abnormalities may result from (a) somatic mutation(s) at a critical time in fetal development or may affect (a) particular organ system(s). Dangers associated with the use of drugs in pregnancy have recently been amplified by the experiences with thalidomide; it should be remembered that aberrations from "normality" in offspring do not necessarily have to be as drastic as those due to thalidomide, e.g., recent unpublished results (J. Werboff, 1962, personal communication) indicate certain behavioral changes in offspring from rats treated during pregnancy with various tranquilizers and other compounds. Thalidomide is the only known sedative whose metabolites are glutamic acid derivatives; those that contain a phthalic acid residue or are derived from the D series are unnatural glutamic acid derivatives, i.e., they do not occur in nature. It is therefore possible that they may interfere with the physiological functions of natural glutamic acid or its derivatives either by taking the place of the latter or by blocking enzymes (oxidases, dehydrogenases, transaminases, etc.). In view of the involvement of glutamic acid in a wide variety of biochemical processes, the embryotoxic effects of thalidomide or its metabolites may be explained by a faulty glutamic acid metabolism.

When thalidomide is administered in an oral dose of 100 mg/kg in the rat, roughly 50% is absorbed and in the dog about 30%; the nonabsorbed portion is excreted unchanged in the feces. Once absorbed, it is excreted in the urine relatively quickly and almost entirely in the form of metabolites. These are (in the dog) products of hydrolysis (and consisting chiefly of glutamic acid derivatives), in contrast to all other sedatives and hypnotics which, as a rule, are inactivated by oxidation, i.e., the introduction of an hydroxyl group, or formation of carbonyl or carboxyl groups (Faigle et al., 1962).

M. Comparative Aspects of Blood Coagulation

While there are numerous studies on a variety of hematologic functions involving cellular elements, hemoglobin, and hematocrit values, comparative aspects of blood coagulation (in man and animals) have been considered but a few times; however a thorough coverage has not been presented before. This seems surprising in view of the fact that in studies of the blood coagulation mechanism, it is often expedient to employ materials obtained from more than one animal species; in such mixed systems, problems of species specificity and species differences are of utmost importance (Didisheim et al., 1959). This review attempts to summarize facts concerning normal and abnormal coagulation in animals; it includes also the most recent advances and research findings on clotting factors in animals. Several of these promise to be of value in both the diagnosis and perhaps therapy of hemorrhagic disorders in man; reference is made particularly to the work on inbred strains of mice performed in the author's laboratory.

Also, an attempt to focus attention on the matter of normal and abnormal blood coagulation in animals seems justified because of the relative lack of extensive investigations; of course, there are many exceptions to this generalization, e.g., the vast amount of work on bovine prothrombin. Certainly there can be little doubt that knowledge of comparative aspects of blood coagulation will benefit laboratory workers concerned with studies on clotting factors in biologic processes. A pertinent example relates to the role of the Hageman factor. Since persons with the Hageman trait never bleed and are detected only by chance (usually prolonged bleeding time), a search for a convenient substitute has been made; indeed, it has been found that this defect is characteristic of the blood of certain vertebrates. In view of this observation comparison with patients suspected of having this trait is now possible.

1. BLOOD COAGULATION IN LOWER VERTEBRATES AND BIRDS

Although only very few investigations concerned aspects of blood clotting in lower vertebrates and birds, most interesting findings have been obtained. Applying the same laboratory criteria as used with mammals for determination of deficiencies in the case of hemorrhagic disorders or increased activities with thrombosis, both extremes in clotting activity have been found, e.g., toads and trout have a very active clotting process while in birds and snakes it is extremely sluggish. Bird's blood is deficient in PTC,* Hageman factor (Didisheim *et al.,* 1959; Thompson *et al.,* 1960), and PTA (Soulier *et al.,* 1959). In the tiger snake, the only clotting factor which, compared to mammalian plasma, appears to be present in adequate amounts is fibrinogen (Fantl, 1961).

Hageman factor activity was tested by the ability of plasmas to influence the clotting of a known Hageman-deficient human plasma; while bird (chicken, goose, turkey, pigeon, muscovy duck, teal) and snake (tiger snake, *Notechis scutatus*) plasma did not correct the deficiency, Hageman factor activation was readily accomplished upon addition of Kaolin suspension and phospholipid to plasma of toad (Cane toad, *Bufo marianas*), trout (*Salmo trutta*), tortoise (long-necked tortoise, *Chelodina longicollis*), lizard (blue tongue, *Tiligua scindoides* and shingle back, *Tiligua rugosa*), opossum (*Trichosurus vulpecala*), and echidna (*Tachyglossus aculeatus*). To produce shortest clotting time the duration of incubation of plasma with Kaolin suspension and phospholipid varied greatly, was less than one minute in the toad, trout, echidna, and tortoise and was several minutes in the lizard (and man); the Kaolin-activated clotting factors (includes both Hageman factor and PTA) were very unstable in the plasma of the toad, trout, tortoise, and echidna.

With regard to thromboplastin formation which is measured by incubation of citrated plasma with phospholipid and calcium ions, several species (lizard, tortoise, and tiger snake) showed considerable reduction.

In measurements of the prothrombin complex determined either by the clotting time in the presence of an homologous organ (lung) extract, Russell's viper venom, and phospholipid or trypsin and phos-

---

* The following names and abbreviations for factors concerned with blood coagulation have been used: antihemophilic globulin (AHF) or factor VIII; accelerator globulin (Ac-G) or proaccelerin or labile factor or factor V; fibrin; fibrinogen; Hageman factor; plasma thromboplastin antecedent (PTA); plasma thromboplastin component (PTC) or Christmas factor or factor IX; proconvertin or serum prothrombin conversion factor (SPCA) or stable factor or factor VII; prothrombin; Stuart-Prower factor or factor X; thromboplastin.

M. Comparative Aspects of Blood Coagulation

pholipid, interesting differences were obtained with respect to both the test species and the test (*in vitro*) system. All lung suspensions, excepting that of the tiger snake, gave the shortest clotting time with homologous plasma. Whereas lizard lung gave almost identical clotting time with lizard as with toad plasma, toad lung showed a distinct species specificity. Also, tiger snake lung suspension gave short clotting times with homologous and lizard plasma while lizard lung was virtually without activity on the prothrombin conversion of the tiger snake. In the presence of both Russell's viper venom and trypsin, tiger snake gave considerably longer clotting times than any other plasma; the clotting time of the trout plasma was shorter in the presence of trypsin than with Russell's viper venom. Since the plasma clotting time as influenced by lung suspension depends upon the activity of prothrombin, factors V, VII, and X, and fibrinogen, a reduction of any could be expected in the tiger snake, a finding similar to birds. While Russell's viper venom time provides information on the activity of prothrombin, factors V and X, and also fibrinogen, trypsin converts prothrombin into thrombin in the absence of any clotting factors; the very slow activation of prothrombin by trypsin in the tiger snake indicated low prothrombin activity. In contrast, in thrombin clotting times and actual measurements of fibrinogen concentrations, the tiger snake showed considerable reactivity, greater than the toad, lizard, echidna, and tortoise and only slightly less than (man and) trout.

These observations (summarized from Fantl, 1961) raise several interesting questions, e.g., relative to the importance of coagulation as an essential process in efficient homeostasis and regarding the relationship between the activity of clotting factors as determined by laboratory tests and clinical syndromes. As for the second question, a direct correlation obviously is not applicable to birds and snakes; they perhaps bypass the plasma coagulation system and rely on vasoconstriction and rapid liberation of tissue thromboplastin following trauma (Bigland and Triantaphyllopoulos, 1960). Regarding the first question the level of fibrinogen only may be the deciding factor in homeostasis of birds and snakes. Although blood coagulation in fish is fundamentally similar to clotting in mammals there is one major difference. It lies in the intrinsic [thrombocyte factor(s)] conversion of prothrombin to thrombin; the extrinsic (tissue factor) systems appear to be more similar, although the participation in the lower species of certain accessory factors such as factors V, VII, and X has not yet been adequately demonstrated (see below). Previous studies on blood coagulation revealed differences in clotting power of blood from various groups of fish. For example, elasmobranch fish have been reported to be essentially hemorrhagic, and it has been suggested that their blood contains little or no prothrombin (Warner *et al.,* 1939). Teleost fish, on the other hand, have a very rapidly clotting blood, although some reports indicate that prothrombin conversion is not always complete (Jara, 1937). In studies on the most primitive group of fish, the cyclostomes, the following findings were obtained (Doolittle and Surgenor, 1962). The plasma of all these fish contained a fibrinogen molecule, capable of being clotted by human thrombin, and a prothrombin molecule, capable of being converted into thrombin, which could hydrolyze p-tosyl-L-arginine methyl ester. The prothrombin activity could be adsorbed on barium sulfate. Accurate assessment of prothrombin conversion factors is confounded by the "species specificity" of protein-protein interactions; thrombocytes play a central role in the intrinsic conversion of prothrombin to thrombin and are responsible for clot retraction (for details see Doolittle and Surgenor, 1962).

2. COAGULATION STUDIES IN MAMMALS

In a study on blood coagulation of various animal species, Didisheim *et al.* (1959) emphasized the great variation in coagulation factors; they also drew attention to differences in results which may be obtained in various test systems, i.e., heterologous versus homologous. Among the species tested were (man), monkey, cow, sheep, dog, cat, rabbit, raccoon, opossum (and chicken and duck).

In tests for prothrombin, proconvertin, and proaccelerin, which were performed according to the usual one-stage procedures employing material from several mammalian species, prothrombin and proconvertin appeared remarkably low in the opossum, and among the species tested man had by far the lowest proaccelerin levels (even lower levels were found in chicken and ducks). Estimation of AHF, PTA, and Hageman factor revealed that AHF activity was very high in cow and sheep plasmas, while the levels of PTC and Hageman factor were rather uniform in all mammalian species (both low or absent in avian

M. Comparative Aspects of Blood Coagulation

species). No very remarkable differences were observed among the species studied in fibrinogen assays. Brain tissue thromboplastin showed a definite species specificity only in the opossum (also chicken and duck) and the thrombin-fibrinogen reaction; opossum fibrinogen was slowly clotted by all thrombins tested, but opossum thrombin clotted only opossum fibrinogen (on the other hand chicken thrombin clotted all of the fibrinogens, but chicken fibrinogen was clotted only by chicken thrombin). Employing homologous systems the opossum (as chickens) showed very low concentrations of prothrombin-proconvertin.

Comparison of relative prothrombin levels among various mammals revealed considerable differences. Assuming a value of 100% for the dog, plasma of man contains 84%, cat 91%, guinea pig 53%, rabbit 89%, and rat 95%. The content in normal dog plasma of prothrombin is approximately 350 units/ml, where a unit is defined as the amount of prothrombin required to form 1 unit of thrombin; the latter is the quantity that will cause clotting of 1 ml of standard fibrinogen solution in 15 seconds under standard conditions. The relative values indicated were obtained by the Quick two-stage test and differ somewhat from those obtained by the one-stage method.

Assays of Ac-G reveal also species differences. Assuming a value of 100% for the dog or 176 units/ml plasma, the following relative levels were obtained: for man 85%, cat 80%, cow 74%, guinea pig 21%, rabbit 94%, and rat 36%. A unit of plasma Ac-G is 1000 times the amount that was present in 1 ml of a reacting mixture of prothrombin, thromboplastin, and calcium and reproduces a standard curve of thrombin production. A variety of specific factor determinations in titered assays have been made for several inbred strains of mice (Meier et al., 1961). No significant differences were found between different strains and sexes of mice in factors affecting blood coagulation (Table VIII). However, three observations deviate from those on clotting of human blood: (1) more rapid loss of prothrombin activity after coagulation as judged by prothrombin consumption tests; (2) low (or variable) platelet activity, either due to low platelet factors or greater resistance of mouse platelets to destruction (in order to obtain consistent clotting of recalcified plasma, admixture of platelet factor reagent is necessary), and (3) a relatively small prothrombin concentration (units of prothrombin/ml about half those of man).

TABLE VIII
SPECIFIC FACTOR TESTS SHOWING COMPARABLE COMPONENT ACTIVITY BETWEEN MOUSE AND HUMAN IN TITERED ASSAYS[a]

Plasma	Factor V		Factor VII		Factor X		Factor VIII		Factor IX		Hageman factor		Prothrombin RV-cephalin	
	PT (sec)	%	PT (sec)	%	PT (sec)	%	PTpln (sec)	%	PTpln (sec)	%	PTpln (sec)	%	RV-cephalin (sec)	%
Mouse	7.8	100	8.3	100	8.2	100	106	100	59	100	106	100	20	100
Control, human	11.2		11.3		11.7		118		59		108		21	
Specific factor-deficient human plasmas + saline	31.0		26		28.8		270		127		240			

KEY: PT = prothrombin time; PTpln = partial thromboplastin (cephalin) time; RV = Russell's viper venom time.

[a] From Meier et al. (1961).

M. Comparative Aspects of Blood Coagulation   143

3. INHIBITORS

Clotting times of whole avian blood were not significantly shorter in glass than in silicone, suggesting that it was devoid of "glass factor." In fact the chicken plasma, upon specific testing for glass factor, when glass-exposed showed no accelerating but actually an inhibiting effect on the silicone plasma. With the possible exception of the Hageman factor assay in which a slight prolongation of clotting time was observed no inhibiting effects could be detected relative to prothrombin, proconvertin, proaccelerin, PTC, and AHF (Didisheim et al., 1959). Specific tests have not been applied for determining the presence of or absence of natural inhibitors in normal mammalian plasma. A thorough investigation of inhibitors had been made in a number of inbred strains of mice; no antithromboplastic activity could be detected (Meier et al., 1961).

4. GENETICALLY DETERMINED CLOTTING ABNORMALITIES DUE TO SPECIFIC FACTORIAL DEFICIENCIES

While a number of species (both avian and mammalian) were found to have single or multiple factorial deficiencies or were even devoid of certain factors relative to the test system employed (mainly human), it is obvious that the hemostatic mechanism is entirely adequate. This accounts for the fact that excessive bleeding does not occur upon vein puncture or minor surgical procedures, e.g., in birds.

In the following subsections, hemorrhagic disorders will be described involving single-factor deficiencies; most of them appear to have a genetic basis (hereditary) or were induced; in the latter instance a genetic (polygenic as well as sex-linked) susceptibility was clearly established.

a. Hageman Factor Deficiency in a Cat

In their studies on coagulation in various animal species, Didisheim et al. (1959) observed very poor thromboplastin generation in one of five cats tested. It could be restored by substitution of either normal human or cat-adsorbed plasma or serum. Also, assays for Hageman factor on three separate occasions showed levels of less than 5% as compared to normal human or cat blood.

Unfortunately, this cat died before arrangements could be made for mating.

b. *Stuart-Prower Factor Deficiency in Inbred Mice*

A spontaneous hemorrhagic diathesis in certain inbred strains of mice has recently been described (Meier *et al.,* 1961). The condition was characterized by bilateral hemothorax with or without multiple bleedings elsewhere (subcutaneous, perivesicular, and subdural) in the body; also the mice were severely anemic and jaundiced. The major histopathologic findings were myocarditis and various degrees of hepatomegaly. Associated with myocarditis was elevation of lactic dehydrogenase; anemia and uterus were thought to be consequent to extensive hemorrhage rather than to hepatic dysfunction since serum phosphatase (alkaline and acid) and transaminase were, except for lowered serum glutamic-pyruvic transaminase, within normal range or only slightly altered.

The mice were found to suffer from single or multiple prothrombin complex deficiencies, especially affecting Stuart-Prower factor (X), PTC (IX), SPCA (VII), and prothrombin (Table IX). Activities of the first three factors as determined in specific factor assays using both human and normal mouse blood were less than 3% and the last was less than 10%; Ac-G (V), Hageman factor, and fibrinogen were normal. Consistently the clotting abnormality which appeared first in the course of the disease was a Stuart-Prower deficiency. Although this defect could be assumed upon the finding of both an abnormal one-stage prothrombin time and TGT (using diluted serum), conclusive evidence rested on the fact that (1) Stuart-Prower factor deficient human serum did not correct the Quick time, prothrombin conversion rate, TGT, or Russell's viper venom (RV) time and (2) a continuous flow curtain electrophoretic fraction obtained from normal mouse serum and containing factor X (Table X) as determined by specific assays was all corrective.

As the disease progressed (eventually fatal to the mice) other factors became involved. A PTC deficiency could be assumed when mouse PTC alone or in admixture with human hemophilia B serum was substituted for reagent PTC; TGT was abnormal. A delayed clotting obtained by the RV and cephalin prothrombin assay, modified so as to

M. Comparative Aspects of Blood Coagulation

TABLE IX
SPECIFIC FACTOR TESTS—PERCENT IN TITERED ASSAY[a]

Plasma	Factor V PT (sec)	Factor V %	Factor VII PT (sec)	Factor VII %	Factor X PT (sec)	Factor X %	Factor VIII PTpln (sec)	Factor VIII %
Test mice[b]	9.8	95(M), 100(H)	16.0	10(M), 17(H)	16.2	3(M), 9(H)	125.0	24(M), 78(H)
Control, mice[c]	7.8	—	8.3	—	8.2	—	106.0	—
Control, human	11.2	—	11.3	—	11.7	—	118.0	—
Def. pl. + saline	31.0	—	26.0	—	28.8	—	270.0	—

TABLE IX (Continued)

Plasma	Factor IX PTpln (sec)	Factor IX %	Hageman factor PTpln (sec)	Hageman factor %	Prothrombin RV-cephalin[d] (sec)	Prothrombin %
Test mice[b]	119.0	3(M,H)	121.0	100(M,H)	30.0	33(M,H)
Control, mice[c]	59.0	—	106.0	—	10.1	—
Control, human	59.0	—	108.0	—	11.7	—
Def. pl. + saline	127.0	—	240.0	—	12.0	—

KEY: PT = prothrombin time; PTpln = partial thromboplastin (cephalin) time; RV = Russell's viper venom time.

[a] Courtesy Dr. John B. Graham, University of North Carolina, Chapel Hill, N. C., vide Meier et al. 1962.
[b] Deficient plasma + test or control plasma = 1:1; SWR/J males, M (mice), H (human).
[c] SWR/J females.
[d] Although the RV-cephalin test measures the combined factor X and prothrombin activity, it was used as a true prothrombin assay in view of the demonstrated factor X deficiency.

include factor X (accomplished by the addition of $BaSO_4$-eluate from normal mouse serum), suggested (1) a prothrombin deficiency or (2) a deficiency of both prothrombin and SPCA. However, direct evidence was obtained in titered assays.

TABLE X

ANALYSIS OF SERUM FRACTIONS OBTAINED BY CONTINUOUS FLOW CURTAIN ELECTROPHORESIS[a]

(0.02 Γ/2; pH 8.6; 825 volts, pooled C57BL/6 serum)

Fraction:	Albumin	Globulins	
R_f values:	0.760	0.576	
Fraction nos.:	22–25	21	17–20
Analysis:		0.31 Mg protein/0.1 ml serum purified 18 times	

Assays
TGT
  Human hemophilia B
  SWR ♂ 434
		PTC (improvement)
		Stuart-Prower factor (corrective)	

SWR ♂ 353	PTC Stuart-Prower factor (improvement)		PTC Stuart-Prower factor (improvement)
RV-Thromboplastin time Stuart-Prower-deficient human plasma + fraction 21		Corrective	
SWR ♂ 373 (proconvertin deficient) + fraction 21		Improvement	
Clotting of recalcified plasma + PF + fraction 21		Corrective	

[a] From Meier et al. (1962b).

In the course of clinico-pathologic studies of the bleeding disorder two observations were particularly noteworthy: (1) a particular strain and sex prevalence, involving SWR/J and DBA/2J males, and (2) the consistency of the lesions both microscopically (heart, liver) and biochemically (liver). While myocarditis was the most severe histologic alteration, liver dysfunction with respect to clotting factor(s) was the most important clinical consequence, being eventually fatal due to exsanguination. These findings stressed the fact that the disease which was acquired rather than due to a mutation, was caused by an agent that acted specifically, inducing well-defined biochemical lesions. Evidence was obtained for implicating ethylene glycol and higher polymers produced by ethylene oxide sterilization of mouse cage bedding. The presence of one or more glycols in ethylene oxide-gassed shavings

TABLE XI
ETHYLENE GLYCOL—TOXICITY STUDIES[a]

Treatment	Coagulation defect[b] total no.		Males affected/ females affected		Clotting defects (%)
	TGT	PT	TGT	PT	
Ethylene oxide-sterilized shavings extract 10^{-1}, 10^{-2}, 10^{-3} dilutions	7/12	8/12	6/1	5/3	62.5
Control shavings extract 10^{-1}, 10^{-2}, 10^{-3} dilutions	3/12	2/12	3/0	1/1	20.8
Ethylene glycol 10^{-2}, 10^{-3}, 10^{-4} dilutions	10/12	10/12	6[c]/4	7/3	84.3
None	2/12	1/12	2/0	1/0	12.5

[a] From Allen et al. (1962b).
[b] Remedied or greatly improved PT and TGT after substitution of purified mouse serum factor X.
[c] One male with hemothorax and jaundice.
[d] Five-tenths milliliter of solution by stomach tube each weekday. (SWR/J males and females 12 weeks old.) Examination for clotting abnormalities each of four Mondays.
KEY: TGT, thromboplastin generation time determination; PT, prothrombin time determinations.

(bedding) was shown chromatographically; also, upon single administration by gavage, 1:5 to 1:10 dilutions of ethylene glycol (Table XI) eliminated or greatly decreased activity of factor X in less than 72 hours (Allen et al., 1962b). In addition, a spontaneous factor X deficiency had been observed only in those bleeding mice that had prolonged contact with sterilized bedding (mice continually chew shavings), and upon abandoning gas sterilization the hemorrhagic disease was eliminated.

No direct explanation was available for the peculiar strain and sex susceptibility; obviously it depended upon a certain genetic composition with the differences in toxicity reactions between strains being both polygenic and sex-linked.

c. *Hemophilia A in Dogs and Swine*

Field *et al.* (1946) reported the occurrence of a condition similar in all respects to human hemophilia A. Beginning usually between 6 and 13 weeks of age, hemophilic puppies showed lameness (originally misdiagnosed as rickets), swelling of joints from hemorrhages, and large subcutaneous hematomata. None of the afflicted dogs lived longer than 7 months and all were males. In eight litters from heterozygous females there were 23 normal males, 17 hemophilic males, and 32 females, all normal. The character was clearly caused by a sex-linked recessive mutation as in man (Hutt *et al.*, 1948).

Female stock from these dogs (heterozygous for the disease) were obtained by Brinkhous, the Medical College of the University of North Carolina. The studies of Graham *et al.* (1949) confirmed that the clotting defect was identical with that found in human hemophilia. The clotting defect was characterized by a prolonged clotting time and a delayed prothrombin utilization, and was corrected by the addition either of thromboplastin or of normal plasma. Repeated transfusions with whole blood or plasma were found to alleviate the hemorrhagic phenomena and permitted growth of affected dogs to maturity essentially free of deformities.

It seems not unreasonable to assume that canine hemophilia occurs more frequently than is indicated by the paucity of reports in the literature. One reason for its being little known may be that owners of purebred dogs are reluctant to incur publicity about any hereditary defects in their stock. Aside from the detailed study summarized above, only two previous recorded instances of hemophilia in dogs could be found. In neither were any tests of the blood performed, but the case histories and clinical findings strongly suggest the possibility. McKinna (1936, quoted by Hutt *et al.*, 1948), observed in the offspring of an Aberdeen terrier bitch three males with bleeding syndromes, one of which died from a large abdominal hematoma at 21 months of age. Merkens (1938, quoted by Hutt *et al.*, 1948) reported a bleeding in greyhounds, with one female producing in four litters eight hemorrhaging males, seven normal males, and three normal females; the afflicted males all died while comparatively young. A third report, by Andreassen (1943), on a bleeding condition in two male Scottish terriers possibly did not represent hemophilia A since one dog lived for 3 years and the other was still alive at 9 years of age.

Earlier, Hogan et al. (1941) discovered hemophilia (-like disease) in some Foundation animals of one strain of Poland China swine and in the Regional Swine Breeding Laboratory at the Missouri station. When inbreeding was practiced, the incidence became greater, with the character expressing itself at about 2 months of age. The "bleeder" hogs had hemorrhages, sometimes fatal, from the gums, sinuses, lips, scratches on nose and ears, and internally, e.g., joints, uterus, and intestines. In this disease, only the homozygote is a bleeder. Since the defect is a simple non-sex-linked recessive (a mating between normal boar and bleeder sow yielding all normal pigs rather than normal females and bleeder males), it differs from that of true human and canine hemophilia which are sex-linked recessives (Bogart and Muhrer, 1942). While the bleeding condition in these swine is from a genetic viewpoint atypical for hemophilia A, it is similar to a case of a woman bleeder (Israels et al., 1951) with a hemophilic father and a mother who was an assured carrier since her father was a bleeder. While this case had been accepted as an authentic example of female hemophilia, especially since the laboratory findings were consistent with a factor VIII deficiency, there is still a certain doubt because of the clinical history. She had no notable hemorrhages until the nineteenth year. Two more cases of similar type have been reported, but one lacked laboratory study (Pinninger and Franks, 1951) and the other represented a direct male to male transmission of the bleeding condition (Merskey, 1950); it is now assumed that the latter might have been a PTA deficiency (Quick, 1959).

d. Hemophilia B in Dogs

The occurrence of a bleeding condition in line-bred cairn terriers which is analogous to the human disorder, hemophilia B or Christmas disease, has been reported (Mustard et al., 1960) in 18 dogs that were available for study. Four dogs had an abnormal clotting mechanism in addition to their history of a bleeding tendency; the latter consisted of excessive hemorrhage following surgery or trauma with formation of subcutaneous hematomata, ecchymoses, bleeding into the anterior chamber of the eyes and from surgical sites. All four dogs were males and either the sons or grandsons of a particular female bleeder. Their clotting times were considerably prolonged, but in the remaining 14 dogs the clotting times were normal. While prothrombin, Russell's viper venom times, and platelets counts were normal in all dogs, prothrombin consumption tests and thromboplastin generation were abnormal. Defective TGT was corrected by the substitution of serum from a normal dog or human, but was not corrected by use of normal platelets or normal $Al(OH)_3$-treated plasma; also whereas admixture of serum from patients with hemophilia A or plasma thromboplastin antecedent (PTA) to bleeding dog's serum corrected for the defect, human hemophilia B serum had no such effect. Since the Russell's viper venom time was normal in these dogs it seems unlikely that the abnormality is similar to a Stuart-Prower factor defect; while in the case of a Hageman factor trait there is no abnormal bleeding, but a very prolonged clotting time and TGT which may be corrected by $Al(OH)_3$-treated normal plasma, involvement of Hageman factor can safely be ruled out. In studies on the effect of platelets on the serum defect it has been found as in man that serum prepared from the clotting of platelet-poor hemophilia B plasma has greater activity in the thromboplastin generation test than serum prepared from platelet-rich plasma. The platelet survival studies showed that in analogy of findings in humans the biological half-life was within the normal range; however, the onset of macroscopic platelet clumping was delayed [similar to the observations by Zucker and Borrelli (1959) and contrary to those by Sharp (1958)].

Therefore, it appears reasonable to conclude that the bleeding disorders in these dogs is genetically and pathologically similar to human hemophilia B. An interesting byline in the paper by Mustard *et al.* relates to the observation that cairn terriers are arthritis susceptible. The possibility that arthritis may result from hemorrhages (hemarthrosis) is suggested since it was implied in the paper that a strain of cairns originating in the British Isles are known bleeders.

e. Factor VII Deficiency in Dogs

An apparently inherited and congenital coagulation defect in beagle dogs similar to human factor VII deficiency has recently been reported; this represents the third congenital clotting defect observed in dogs (Mustard et al., 1962a). None of the affected dogs had been known to have a bleeding tendency; they were apparently healthy animals with no clinical or laboratory evidence of liver disease. The dogs which were accidentally discovered to have the clotting defect belonged to an inbred strain of beagles maintained at the research unit of the Ontario Veterinary College. Construction of a family tree indicated that the defect was inherited and affected both male and female dogs. All affected dogs (four females and two males) studied had been found to have a normal clotting time, platelet count, platelet clumping, prothrombin consumption test, Russell's viper venom time, and TGT; however, prothrombin times were prolonged. Addition to the plasma of affected dogs of 10% normal dog or human serum, but not normal $Al(OH)_3$-treated dog or human plasma from patients on dicoumarol therapy, shortened the prolonged prothrombin time. A factor VII deficiency was shown directly in activity assays using plasma from a human patient who had been receiving dicoumarol for a period of 3 days. Liver dysfunction* with which factor VII deficiency is often associated

---

* Factor VII deficiency readily occurs in association with impaired liver function in man (Alexander et al., 1959) and animals. Clotting factor determinations in mice were found to be extremely sensitive indicators of liver function, e.g., abnormalities associated with a variety of bacterial and other diseases showed up even prior to the detection of other functional deficiencies. A pertinent example relating to factor X deficiency induced by ethylene glycol has been described. While in that instance hepatotoxicity was specific for factor X, most liver damage usually affects several clotting factors, e.g., hypoprothrombinemia is usually most severe in chloroform poisoning in the dog but labil factor is also greatly diminished.

The relationship between vitamin K_1 and "prothrombin complex" components is well established. It seems relevant, however, to mention the fact that this relationship depends upon the animal species. In studies on the etiology and epidemiology of a spontaneous hemorrhagic diathesis in mice (see above) various deficiency diets and feed containing an excess of soybean meal were used; of considerable interest was the fact that total lack of vitamin K-complex did not induce a factor VII deficiency and bears upon other evidence that under ordinary conditions intestinal microorganisms are able to supply a considerable amount, if not all, of the required amounts of vitamin K. However, it has been observed that certain clinical signs may result in mice from a dietary tocopherol deficiency and that they are sex-dependent; while male mice grow equally well with

was ruled out by a variety of tests (cephalin cholesterol flocculation, bromsulfophthalein retention, serum protein, and alkaline phosphatases); also the parenteral administration of vitamin K_1-oxide was not found to change the prothrombin values during the following 36 hours.

While bleeding tendency in human factor VII deficiency occurs at factor VII levels of 10% or less (easily bruised skin, purpura, and ecchymosis), skin hemorrhages are not easily detected in dogs; also the deficiency in the affected dogs, from the values factor VII assays, does not appear to be sufficiently low to cause spontaneous bleeding.

f. Multiple (Pre-) Coagulant Factor Deficiencies (Left Auricular Thrombosis) in Mice

Left auricular thrombosis occurred in high incidence (66%) among older breeding females or inactive breeders of BALB/cJ. The significance of repeated pregnancies (and not age alone) was indicated by the fact that BALB/cJ males and virgin or ovariectomized females had never been found to suffer from the condition. While pregnancies in most strains did not affect coagulation or only irregularly caused some activity reduction in certain precoagulant (stage I) factors, BALB/cJ females suffered from severe deficiencies of PTC (IX; about 40% of normal), AHF (XIII; about 60% of normal), Stuart-Prower factor (X, about 50% of normal), and prothrombin (about 33% of normal)

---

or without α-tocopherol, females lag somewhat in growth. Deficiency may manifest itself also in deaths and resorption of embryos during the later stages of the gestation period; and after prolonged deprivation signs of muscular dystrophy or deaths ensue. Similar signs of vitamin K deficiency in female mice may be induced by *dl*-α-tocopherolquinone, a vitamin K analog. Hemorrhages occur that are strictly confined to the reproductive system and in pregnant mice only; the compound is without effect in males and non-pregnant females. However, pregnant mice that received daily oral doses of 100 mg of the quinone had similar prothrombin times as the controls. 3,3-Methylene-bis-4-hydroxycoumarin, which caused some signs of vitamin K deficiency (that were reversed by vitamin K), did not produce resorption of fetuses and vaginal hemorrhages in pregnant mice.

This compound, interestingly enough, is the toxic principle of spoiled sweet clover hay that had caused a rather troublesome and puzzling bleeding disease in cattle of North Dakota and various sections of Canada.

Quick found that prothrombin dropped abruptly 24 hours after feeding the toxic hay; when the animals were transfused with normal citrated plasma an immediate rise in the prothrombin concentration occurred. Also, upon feeding alfalfa, which is rich in vitamin K, together with the toxic hay the prothrombin failed to decrease. This was the first experimental evidence of the antagonistic action of the principle in spoiled sweet hay.

occurring shortly before parturition; Hageman factor and Ac-G(V) were always normal (Table XIIa). These prothrombin-complex abnormalities disappeared a few days *post partum* (Table XIIb); plasma rebound might be 20 to 25% above normal 2 days *post partum*. The

TABLE XIIa
SPECIFIC FACTOR TESTS: PREPARTUM

Plasma	Factor X		Factor VIII		Factor IX		Prothrombin	
	PT	%	PTpln	%	PTpln	%	RV Cephalin	%
BALB/c males	7.9	>100	108	>100	59	100	20	100
Control, human	11.7		118		59		21	
Deficient plasma + saline	30.8		261		131		—	
BALB/c females	19.8	~50	159	~60	101	~40	52	~33

KEY: PT = Prothrombin time; PTpln = partial thromboplastin time; RV Cephalin = Russell's venom-cephalin prothrombin assay.

TABLE XIIb
RELATIVE PROTHROMBIN LEVELS: POST PARTUM[a]

Plasma[b]	Prothrombin (units/ml.)	
	Range	Mean
BALB/c females	190–211	198
Non-pregnant controls	162–181	169

[a] From Meier and Hoag (1962d).
[b] 10 animals each.

rebound phase was only moderately or not reflected in the one-stage prothrombin time. Attempts to measure also fibrinolytic activity in blood from thrombotic mice failed because of the very low activity found normally in mouse blood (Meier *et al.*, 1961).

It was felt that the postparturient prothrombin rebound, as observed in man following dicoumarol and heparin therapy, might have greatest clinical impact. The thrombus probably formed rather quickly and was primary while endocardial damage was most likely secondary. The clinical condition associated with advanced stages of auricular thrombosis consisted of lack of vigor, labored breathing from lung congestion and in most instances subcutaneous edema; at necropsy the left auricle was 3 to 5 times the normal size. Since the disease can be diagnosed readily or at least suspected (particularly in breeders that were pregnant more than 6 times), the condition lends itself for study of thrombosis formation (hormonal imbalance) and lysis or prevention and screening of potential fibrinolytic compounds (Meier and Hoag, 1962d).

A similar condition, but induced, occurs when young adult mice of C, C57BL/6, DBA, and Swiss stocks are fed a hyperlipotropic diet containing (28%) fat as lard and (8%) protein as casein. At about 7 weeks of feeding such diets, all stocks develop atrial necrosis resulting in formation of mural thrombi, reaching critical or terminal dimensions in 10 to 12 weeks. Betaine (hydrochloride, 2 mg/100 gm of diet) does not prevent the lesions (Ball, 1962).

5. Therapy of Clotting Disorders

There are conflicting theories on the management of hemophilic hemorrhages in man; this is particularly pertinent relative to the use and effectiveness of serum and plasma transfusions in correcting the clotting defect in hemophilia B. However there is evidence that (a fraction of) normal human, beef, pig, or mouse plasma will correct the clotting defect of hemophilic blood *in vitro,* and will establish normal hemostasis for a time after its injection *in vivo.*

The experience in man is that in the case of hemophilia A, therapy with blood and plasma (fresh citrated, heparinized, or fresh lyophilized plasma is usually successful.

Similarly, good to excellent results have been obtained with concentrated AHF preparations of both human and animal (beef, pig) origin (van Creveld and Mochtar, 1959). The use of concentrated AHF and its effectiveness in emergencies is particularly relevant especially if it is of heterologous (animal) origin. Bidwell (1955) has produced ox and pig AHF which is up to 8000 times as active as the same weight of fresh human plasma protein (if AHF level is above 30% and provided there is no other hemostatic defect, no bleedings occur). Clearly, the disadvantage of animal AHF is its antigenicity in man. However, it is no more of a deterrent in emergency use than is the antigenicity of animal antitoxins used in the treatment and prophylaxis of diphtheria, tetanus, and gas gangrene. It only means that a particular animal AHF can be used safely for only one course of treatment (lasting about 10 days) and should therefore be reserved for serious situations only. It may be possible that in the future more pure and perhaps less antigenic preparations can be obtained (Macfarlane and Biggs, 1959). In dogs afflicted with hemophilia A, repeated transfusions with whole blood and plasma from normal dogs alleviated not only hemorrhagic phenomena but also allowed them to grow to maturity free from deformities appearing in untreated animals (Graham et al., 1949).

Regarding the use of serum and plasma in patients with hemophilia B, Welsh (1960) reported that serum was more effective while Nour-Eldin and Wilkinson (1958) were of the opinion that plasma transfusions were superior to serum. In addition, there have been different views about the stability of factor IX in plasma; these seemed to be related to variations in the methods of plasma collection, storage, and assay for factor IX (Geratz and Graham, 1960). In a study on the relative effectiveness of serum and plasma transfusions in correcting the clotting defect in canine hemophilia B, it was shown that both are equally good (Mustard et al., 1962b). The increase in factor IX activity in the transfused animals was proportional to the activity of the transfused material; factor IX activity of the transfused material was considered an important factor in determining the response of the recipient. Changes in factor IX were determined by the TGT, which is the most sensitive method, by platelet clumping time, and by clotting time. It was thought significant that platelet clumping time was shortened upon transfusion and correction of factor IX deficiency, in view of the concept that factor IX may be involved in the development of platelet viscous metamorphosis.

Research on clotting factors from inbred strains of mice has been carried far enough to separate (Allen et al., 1962a) and purify (Meier et al., 1962b) a number of them. In view of the observations made (alluded to in other sections) it was possible to prepare test reagents and single or multiple factor-deficient plasmas. Since factors V and VII can be obtained essentially free from contamination of other factors and if prepared in sufficient quantities they could be used therapeutically (in man) or for positive affirmation of deficiencies in test sera (correction of defects *in vitro*).

6. Purification, Properties, and Composition of Bovine Prothrombin

Extensive work on the *purification* of bovine (and human) prothrombin yielded products of sufficient purity for study including partial physiochemical characterization (Miller and Seegers, 1950). The most successful method with respect to preparation of prothrombin consisted of employing adsorbing agents, i.e., $Mg(OH)_2$; fractionation of plasma with the use of alcohol, ether, or salts had not yielded fractions that contained prothrombin in sufficient purity to be of value for studies of its properties. Even after prothrombin of high purity could be obtained by adsorption techniques, further purification was difficult although the specific activity could be increased by adsorbing impurities on $BaSO_4$.

Comparison of the *properties* of bovine and human prothrombins in regard to criteria such as electrophoretic behavior, autocatalytic activation and its inhibition; activation with thromboplastin, platelets, and calcium; UV adsorption, and stability characteristics revealed identical reactions (Laney and Waugh, 1953); the only difference found relates to the fact that bovine prothrombin preparations of comparable specific activity were homogeneous in the ultracentrifuge while the human prothrombin was not (Seegers and Alkjaersig, 1953).

Purified bovine prothrombin was found to be antigenic for both guinea pigs and rabbits. Rabbit immune serum possessed precipitating antibodies capable of specific reaction with it, prothrombin activity being removed coincidentally with this specific precipitation. Therefore, it can be assumed that bovine prothrombin itself possesses antigenic properties. In addition bovine prothrombin preparations had been reported to have an approximate molecular weight of 62,700 and as being a glycoprotein; both of these requirements would suit the requirements of an antigen (Halick and Seegers, 1956).

In the preparations used for immunologic studies a second antigen component was observed by means of both precipitation and sugar diffusion techniques and was thought to be possibly Ac-G which was known to be present in these same preparations. Importantly, however, it could be shown that antibodies specific for prothrombin may be produced in laboratory animals; immunologic techniques can now be added to the armamentarium for the study of blood clotting mechanisms.

The fact that it had been possible to isolate prothrombin in relatively pure form made feasible analyses of its *amino acid composition* (Laki *et al.,* 1954). Such studies were of particular interest and were thought to serve as a basis for an understanding of the mechanism involved in the formation of thrombin from prothrombin. Thrombin is obtained by activation of prothrombin in concentrated sodium citrate solution, a process that results in loss of a large fraction of carbohydrate and nitrogen. The full amino acid composition of bovine prothrombin was measured by using ion-exchange chromatography. Certain of the amino acids present were measured analytically; eighteen amino acids and hexosamine were found, glutamic acid, aspartic acid, and arginine being present in highest weight percentage. No resemblance could be detected between serum albumin and prothrombin, although the two are closely related in respect to many other physicochemical characteristics.

7. Distribution of Coagulation Proteins in Normal Mouse Plasma

A successful approach to the study of clotting factors and their hemostatic function was the assay of fractions from normal mouse serum obtained by continuous flow curtain electrophoresis in various deficiency systems. Since in mice with hemorrhagic diathesis, a deficiency of factor X appeared earliest in the course of the disease or greatly increased its severity, a Stuart-Prower factor fraction from mouse serum was first prepared. This fraction, corresponding to the leading edge of α_1-globulin of serum protein, was rich in factor X with very little contamination by plasma thromboplastin component. Although X-deficient human plasma does not correct for Quick time, prothrombin conversion rate, thromboplastin generation time, or Russell venom time, the mouse serum fraction was all corrective. The technique of continuous flow curtain electrophoresis was applied for studying the distribution of various coagulation factors in normal C57BL/6J mouse plasma. Clotting factors were identified by standard and specialized assays (substitution analyses by employing known single-factor-deficient human plasma). In controls, saline was added to deficient human plasma (1:1) in place of curtain fractions. Presence of activity was indicated when either prothrombin time or thromboplastin generation time was improved upon admixture of curtain fractions over that of the controls. Although separation of coagulation proteins in human plasma is incomplete, complete separation being effected only by utilization of serum rather than plasma or of specific

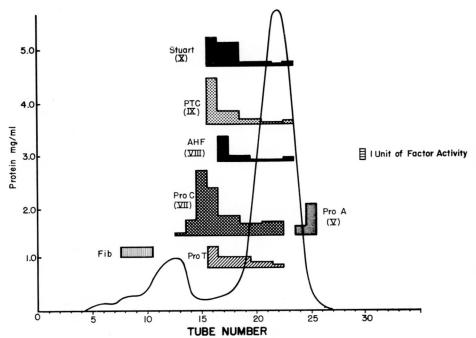

GRAPH V. Distribution of clotting factors in C57BL/6J mouse plasma, expressed in units of relative activity. Plasma separation carried out in Spinco continuous flow electrophoresis apparatus using Veronal buffer $\Gamma/2$ 0.020, pH 8.6 at 52 Ma, 720 volts. Temperature was held at 11° C. (From Allen *et al.*, 1962a.)

factor-deficient plasma, the results of analyses of mouse plasma fractionation indicate that factors V and VII may be obtained free from other factor contaminations. In tube ranges with overlapping factors, certain of them may be eliminated by $BaSO_4$-adsorption, with or without subsequent citrate elution.

The genetic homogeneity of this inbred strain of mice with respect to clotting factor mobilities provides for excellent reproducibility of the fractionation procedure even when pooled plasma samples are used. In another separation a similar distribution pattern was obtained. The distribution of mouse coagulation factors was found to be similar to that of man in respect to the familiar γ-, β-, and α-globulin, and albumin pattern. However, the zones of activity were narrower and the boundaries of activity more distinct in the mouse plasma (Graph V). These observations suggest the possibility of using mouse plasma or serum for the preparation of test reagents and single- or multiple-factor-deficient plasmas.

8. Separation and Purification of Clotting Factors from Inbred Mice

Analysis of blood clotting factors performed on fractions obtained by continuous flow curtain electrophoresis of mouse plasma indicated that factors V (completely) and VII (in part) can be obtained free from other factor contamination; these may be concentrated by dialysis against 10% gelatin and further purified by a variety of techniques applied to the dialyzate. In view of previous work demonstrating that one such curtain electrophoretic fraction from C57BL/6J serum contained both factor IX (PTC) and X (Stuart), attempts were made to separate and purify the two. This subfraction representing a 242-fold purification of factor X activity compared to normal serum contained 0.31 mg protein/0.1 ml and corresponded to the leading edge of the α-globulin zone under the conditions employed. Admixture (1:1) to both X-deficient human plasma and plasma obtained from mice early in the course of a spontaneous hemorrhagic diathesis or fed (by gavage administration) ethylene glycol which inhibits factor X activity, corrected for all clotting abnormality and also greatly improved TGT in substitution with equal parts of human hemophilia B serum.

Starch gel electrophoresis was used for further purification of these two factors. The protein content of each eluate was assayed for determinations of specific activity. Recovery of factor X was 55.6% and of factor IX about 25%, losses of activity occurring chiefly from denaturation and also insufficient elution (retention by gel after centrifugation). Although complete separation of the two factors was not quite achieved, the degree of factor enrichment could be estimated from the percent activity per milligram of protein, e.g., specific activity of factor X for normal C57BL/6J serum was 1.25% per milligram of protein while that of eluate No. 2 (containing factor X) was 6000% per milligram. Therefore, calculated for a solution of serum or plasma containing 100% factor X activity per milliliter, the protein concentration is less than 10 μg per milliliter (or 1 mg%). This degree of purification is 2–4 times greater than the best achieved previously with human blood (plasma concentrate) by either starch gel electrophoresis (under conditions slightly different from ours) or cellulose column chromatography (Meier et al., 1963).

Of considerable interest was the finding of three different electrophoretic species (which may be comparable to iso- or hybrid enzymes) of factor IX (α_1 and α_2 regions). Since the range of activities of

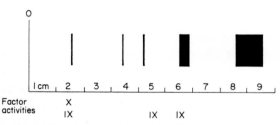

GRAPH VI. Localization of clotting factor activity by assay following elution. Starch gel electrophoretic species of factors IX and X; activities were determined by substitution analyses. (From Meier et al., 1963.)

factor IX and X following plasma continuous flow curtain electrophoresis was identical (tubes Nos. 15–24), positive assays (substitutions analyses) for both in the slow α_2 was not unexpected, but it is conceivable that the two factors may through interactions between them form a complex. The two additional peaks of factor IX activity were not anticipated and have not previously been noted (Graph VI). Preliminary observations of immunoelectrophoresis indicate that factor X (or the complex) migrates with or is an α_2-lipoprotein (α_2-III globulin).

9. FIBRINOLYTIC ACTIVITY OF MOUSE ENDOMETRIAL SECRETIONS (UTERONE)

Ligation of the uterine cervix in mice results in a certain proportion of animals in the accumulation of uterine fluid (uterone) without any inflammatory change in the uterine wall (Homburger et al., 1955). The rate of accumulation of hydrouterine fluid is dependent upon endocrine factors (Homburger and Tregier, 1957). The biologic activities of uterone (increase in uterus size and decrease of adrenals in castrates; inhibition of accelerated growth of transplanted tumors following castration and hysterectomy; and hemorrhagic-necrotizing ophthalmia) are characterized by a certain degree of strain specificity: for example, the uteri of castrates were enlarged only in BALB/c, AKS, and Swiss strain, but not C57BR, and occular hemorrhagenicity ("black-eye syndrome") occurred in BBF_1, BALB/c, and AKS mice, but not Swiss, when uterone was injected intraperitoneally (Homburger et al., 1957). The hemorrhagenic factor has now been separated into two components (by dialysis): one seems to have heparin-like action but differs from heparin regarding its electrophoretic mobilities and relative activity; the second is fibrinolysin-like (Bernfeldt, Homburger, and Bang, 1962, personal communication). The latter seems of utmost interest because of its extreme potency (in standardized assays) to lyse fibrinogen. Indications are that protein breakdown in uterone liberates an activator of (a) proteolytic enzyme(s) that lyses fibrinogen (Bang, 1962, personal communication). Present and future biochemical investigations are directed toward isolation and analysis of this factor; a new type of fibrinolytic agent of high biological activity may be developed.

Procedures for obtaining uterone from other species have not been successful. In mice 5 to 20 ml/uterus may be harvested; it may be lyophilized or stored frozen. Injections of crystalline ovonuecoid in doses of up to 128 mg/mouse did not produce the "black-eye syndrome"; however, mice receiving 32 mg of dry uterone intraperitoneally develop it in 16 to 72 hours.

10. Endocrine Factors Influencing Fibrinolytic Activity and Integrity of Vascular System

Stress, excitement, and physical activity are known to alter fibrinolytic activity (Biggs et al., 1947); fibrin deposition and plasma fibrinolytic activity have repeatedly been considered as important factors in the promotion or prevention of coronary thrombosis. As a possible basis for the thrombotic episodes in *coronary disease of rats,* in relation to sex differences, a systematic study on the gonadal influences on plasma fibrin and fibrinolytic activity is particularly revealing. Plasma fibrin is statistically significantly higher in normal and gonadectomized males than females. Fibrinolytic activity is the same in males and females under basal conditions, but is depressed after orchiectomy; it is especially low 3 hours after a fatty meal, but only in normal males (Gillman and Naidoo, 1958).

Evidence that pregnancies may influence the incidence of vascular lesions was derived also from studies in rats; Wexler and Miller (1958) noted that breeding females treated with corticotropin were more susceptible to severe arteriosclerosis than non-breeding females or males. In studies on the spontaneous incidence of vascular injuries in breeding and non-breeding male and female stock rats on a standard diet, male rats were almost devoid of aortic and coronary mineralization, while the incidence was high in females and more frequent in breeders than non-breeders (Gillman and Hathorn, 1959). Mineralization (calcium and/or iron coating) of the internal elastic membranes of coronaries, mesenteric and uterine arteries was most common. Calcification, with heavy mucopolysaccharide accumulations and cartilage formation, occurred in the myocardium; also patch areas of increased myocardial reticulin. These lesions are similar to those described for healing arteries after short-term toxic doses of calciferol; also there was a close association between mucopolysaccharide accumulations, mineral deposition, and fibrosis (Gillman and Gilbert, 1956). Selye (1958) has shown that corticoids markedly increase the susceptibility of the coronary arteries in rats to similar traumata; the corticotropin treatment applied by Wexler and Miller (1958) may thus have aggravated pre-existing cardiovascular lesions in their breeding females.

In a survey of heart lesions in retired breeder mice, dystrophic calcification of myocardium was found in almost all mice of strains C3H, C3HeB, DBA/1, and DBA/2; less involvement was found in BALB/c and A and very little involvement occurred in C57BL/6, C57BL/10, C57L, RIII, 129, MA, and SWR. Damage was more extensive at younger ages in breeding females, except DBA/2, where males also were severely affected. Virgin females were relatively free of lesions. Left auricular thrombosis and coronary arteriosclerosis occurred in BALB/c and coronary arteriosclerosis of the SWR strain (Hummel, 1962). While the fibrinolytic activity and plasma fibrin appear to be involved in the pathogenesis of arteriosclerosis of rats, no clues are available for the lesions in the SWR data on left auricular thrombosis in BALB/c have been presented (see above).

IV. Epilogue

Experimental animals represent the integrating link between basic information from lower organisms (much is already at hand, e.g., in microbes) and man. Obviously, insight into the nature of relationships between units of genetic transmission and their intracellular products may be traced advantageously in genetically controlled hosts. Reasons have been cited and illustrative examples of the importance of genetic quality control given. *It is hoped that greater attention will be turned toward the experimental hosts;* since species and strain specificities of various reactions do occur, the best systems in animal experimentation should be carefully scrutinized.

In man, any approach to the problem of the nature of gene function is necessarily rather indirect. In practice, what is observed in the first place is some difference between people to a particular kind of character; by analysis of the manner in which this character is distributed in a population, one may be able to infer the existence of more than one allelic gene at some chromosome locus. A detailed investigation of the biochemical pathology of individuals presumed to carry the different allelic genes in their various combinations may then enable one to obtain some idea of the primary and specific manner in which they differ from one another metabolically. Generally, the character or phenotypic difference serves as a starting point. The sequence of causal relationships connecting them is usually complex, involving phenomena at many different levels of biochemical and physiological functioning. Genetic and biochemical analysis of the conditions termed "inborn errors of metabolism" lead to the conclusion that a more or less direct relation exists between the presence of a certain gene and the absence of a particular enzyme. Examples of such conditions with the respective enzyme deficiency in man are listed (Table XIII); additional enzyme (and substrate) deficiencies concern the blood clotting mechanism. Conditions analogous to man occur in animals. Other genetic differences relate to the structures of macromolecules such as proteins and mucopolysaccharides, composition of body fluids, secretions, and excretory products. Of the example discussed, the finding of elevated α_2-globulin in prediabetic Chinese hamsters is of great interest; it may be that there is not only an elevation of normally occurring protein but also of abnormal complexes. Ultimately the significant differences probably lie in the structural organization of nucleic acids present in the cell nucleus.

TABLE XIII
ENZYME DEFICIENCIES IN INBORN ERRORS OF METABOLISM[a]

Condition	Deficient enzyme
Phenylketonuria	L-Phenylalanine hydroxylase
Alkaptonuria	Homogentisic acid oxidase
Tyrosinosis	p-Hydroxyphenylpyruvic acid oxidase
Albinism	Tyrosinase
Cystathioninuria	Cystathionine cleavage enzyme
Glycogen storage disease (a)	Glucose-6-phosphatase
Glycogen storage disease (b)	Amylo-1,6-glucosidase
Glycogen storage disease (c)	Amylo-(1,4–1,6)-transglucosidase
Galactosemia	Galactose-1-phosphate uridyl transferase
Methemoglobinemia (one type)	Diaphorase I (methemoglobin reductase)
Congenital hyperbilirubinemia	Glucuronyl transferase
Acatalasemia	Catalase
Hypophosphatasia	Alkaline phosphatase
Goitrous cretinism	Dehalogenase

[a] From Harris (1959).

Obviously, to unravel some of the mysteries still to be resolved, pharmacogenic studies may be anticipated to produce exciting new discoveries. A wealth of experimental material is now available for intensive investigation. In this respect, the studies on the ribonucleic acid (RNA)-induced biosynthesis of specific proteins are of interest. The type of specific protein synthesized by the RNA recipient cells relates directly to the tissue source of RNA; thus, liver RNA-treated Nelson ascites, and Novikoff hepatoma cells acquire the biochemical specificity of producing such liver specific proteins and enzymes as serum albumin, tryptophan pyrrolase, and glucose-6-phosphatase (Niu et al., 1962). The application of polynucleotides and nucleoproteins for replacement of faulty cellular macromolecules may be the next step in the control of metabolic errors. An alternate approach lies in influencing enzyme activities by exogenous means. In genetically con-

IV. Epilogue

trolled defects, metabolic adaptation and induction, e.g., by a corresponding enzyme substrate or related inductor, seems of limited potential, since obviously the code for a particular enzyme is lacking. However, in instances where a genetically determined inhibitor is responsible for inactivation or binding of an enzyme or hormone, its biologic effects could be successfully interfered with; an example given is that of the inhibition of phenylalanine hydroxylase in dilute-lethal mice. Also, if a binder-protein causes insulin to be masked either liberation of insulin may be induced pharmacologically or the action and synthesis of the binder blocked. The principle of approach is comparable to metabolic autogonism and inhibition known especially in cancer chemotherapy. Another postulate relates to restitution and replacement of insufficiencies; it seems questionable whether or not cells suffering from such a deficiency could be "nursed back" to a more normal state by providing them with holo-, apo-, or co-enzymes having specific activity.

A threshold of many new developments has been reached in the field of pharmacogenetics. It is obvious that pharmacologic studies may be profitably applied to the analysis of mammalian gene-action. In order to understand the physiologic mechanism in response to drugs, differences between inbred strains of animals may be of help. Special purpose mutant stocks are of particular interest since existing differences are due to single pairs of genes. These may be propagated either by (1) inbreeding with forced heterozygosis or (2) repeated crossing of the mutant animal to a standard inbred strain. Both systems result in two kinds of animals, normal and mutant, appearing in the same litters and as genetically alike as possible, except for the one locus which has been forced to segregate in the cross. Drugs may not only be influenced by a gene in their action, toxicity and metabolism, but they may in fact be responsible for the discovery of certain genes. Pertinent examples in man relate to primaquine sensitivity due to a deficiency of glucose-6-phosphate dehydrogenase and suxamethonium sensitivity which is due to a lack of serum cholinesterase (pseudocholinesterase). As with the control and formation of enzymes, a variety of inherited diseases result in peculiarities in the synthesis of proteins or protein groups, e.g., the failure to synthesize γ-globulin or the production of abnormal hemoglobin (sickle-cell anemia, etc.). Two very intriguing syndromes, one characterized by the complete absence of serum β-lipoprotein or a β-lipoproteinemia (Salt *et al.*, 1960) and the other by the absence of α-lipoprotein or Tangier disease (Fredrickson *et al.*, 1961) may furnish valuable information regarding the pathogenesis of atherosclerosis since one of the feasible chemotherapeutic approaches may be the specific inhibition of the protein moiety of lipoprotein.

Appendix: Procurement of Animals for Research

Since the nature of many scientific purposes has created special requirements for animals, problems of procurement are raised. Specific inquiries regarding availability of particular types of animals may be directed to fellow scientists who maintain colonies of them and have published information regarding them. In addition, research workers are assisted in their quest for suitable animal supply by a catalog of commercial sources prepared by the Institute of Laboratory Animal Resources (ILAR).* This catalog lists sources for the common domestic species and all commercially available animals in phylogenetic order, beginning with Protozoa and ending with Chordata. A continuing inventory of commercial laboratory animal stocks provides the Institute, and through it research workers and commercial producers, valuable information on the current inventory of experimental material and the trends in research animal utilization. Also, as part of the ILAR's information service it analyzes and answers inquiries regarding specialized needs for scientific programs.

In choosing the various types of mice for research, valuable information may be obtained from a "Handbook on Genetically Standardized JAX Mice," which is available through the Production Department of the Roscoe B. Jackson Memorial Laboratory; it refers also to various other services, e.g., bibliographic, available from the Laboratory. The Jackson Laboratory maintains over 60 inbred strains of mice, produces six kinds of F_1-hybrids (this number can be increased whenever other hybrids become useful in research) and propagates over 130 named mutant genes. Nearly all of these mutant mice are available in small numbers or as breeding pairs; a few mutant stocks are presently in sufficient demand to justify maintenance in larger numbers, but other stocks may be added if warranted. Prices for mutants depend of course on the ease with which they can be bred; those with reasonable

---

* Institute of Laboratory Animal Resources (ILAR), 2101 Constitution Avenue, Washington 25, D. C.: Animals for Research, A Catalogue of Commercial Sources, 1959.

viability and fertility cost less than those that are either inviable or infertile and must be bred from tested heterozygotes or from females bearing transplanted ovaries.

References

Adams, R. D., Denny-Brown, D., and Pearson, C. M. (1962) "Diseases of Muscle: A Study in Pathology," 2nd ed., p. 340. Harper, New York.

Adamson, R. H., and Black, R. (1959) Volitional drinking and avoidance learning in the white rat. *J. Comp. Physiol. Psychol.* **52**: 734-736.

Adrouny, G. A., and Russell, J. A. (1956) Effects of growth hormone and nutritional status on cardiac glycogen in the rat. *Endocrinology* **59**: 241-251.

Akamatsu, S., Kiyomoto, A., Harigaya, S., and Ohshima, S. (1961) Inhibition of beta-glucuronidase by oral administration of d-glucosacchoro-1,4-lactone. *Nature* **191**: 1298-1299.

Alexander, B., Kliman, A., Colman, R., Scholtz, E., and Di Francesco, A. (1959) New "Hemophiliod" defects: some clinico-laboratory and experimental abnormalities in thromboplastin generation. *In*: "Hemophilia and Other Hemorrhagic States" K. Brinkhous and P. DeNichola, eds.), pp. 137-157. Univ. of North Carolina Press, Chapel Hill, North Carolina.

Allen, L., Burian, H. M., and Braley, A. E. (1955) A new concept of the development of the anterior chamber angle. *A.M.A. Arch. Ophthalmol.* **53**: 783-798.

Allen, R. C., Meier, H., and Hoag, W. G. (1962a) Distribution of coagulation proteins in normal mouse plasma. *Science* **135**: 103-104.

Allen, R. C., Meier, H., and Hoag, W. G. (1962b) Ethylene glycol produced by ethylene oxide sterilization and its effect on blood clotting factors in an inbred strain of mice. *Nature* **193**: 387-388.

Allison, A. C., Rees, W., and Burn, G. P. (1957) Genetically controlled differences in catalase activity of dog erythrocytes. *Nature* **180**: 649-650.

Altman, K. I., Russell, E. S., Salomon, K., and Scott, J. K. (1953) Chemopathology of hemoglobin synthesis in mice with a hereditary anemia. *Federation Proc.* **12**: 168.

Ambrus, J. L., Guth, P. S., Goldstein, S., Goldberg, M. E., and Harrison, J. W. E. (1955) Toxicity of histamine and antagonism between histamine and anti-histamines in various strains of mice. *Proc. Soc. Exptl. Biol. Med.* **88**: 457-459.

Amin, A., Chai, C. K., and Reineke, E. P. (1957) Differences in thyroid activity of several strains of mice and F_1-hybrids. *Am. J. Physiol.* **191**: 34-36.

Ammon, R., and Werz, G. (1959) Ueber die verteilung der atropin- and cocain-esterase in der Leberzelle des Kaninchens. *Z. Physiol. Chem.* **314**: 194-197.

Andreassen, M. (1943) Studies on the thymolytriphatic system. *Acta Pathol. et Microbiol. Scand. Suppl.* **49**: 1-160.

Ashton, G. C., and Braden, A. W. H. (1961) Serum beta-globulin polymorphism in mice. *Australian J. Biol. Sci.* **14**: 248-258.

Assali, N. S., and Suyemoto, R. (1954) Studies on toxemia of pregnancy; effects of corticotropic hormone (ACTH) on hemodynamics and excretion of electrolytes of normal pregnant women. *Metabolism* **3**: 303-312.

Axelrod, J., and Laroche, M. J. (1959) Inhibitor of O-methylation of epinephrine and norepinephrine *in vitro* and *in vivo*. *Science* **130**: 800.

Axelrod, J., and Tomchick, R. (1960) Increased rate of metabolism of epinephrine and norepinephrine by sympathomimetic amines. *J. Pharmacol. Exptl. Therap.* **130**: 367-369.

Babel, J. (1944) Etude anatomique de l'hydrophthalmia familiale et hereditaire du lapin. *Arch. Sci. Phys. Nat.* **26**: 100-103.

Babikian, L. G. (1962) Effect of hydrocortisone, corticosterone and cortisone on visceral and total body fat in albino mice. *Physiol. Zool.* **35**: 314-322.

Ball, C. R. (1962) Cardiac lesions in mice fed high fat, low protein diets. *Anat. Record* **142**: 212.

Barrows, C. H., Jr., and Roeder, L. M. (1962) Age differences in concentrations of various enzymes in tissues of rats. In "Biological Aspects of Aging" (N. W. Shock, ed.) pp. 290-295. Columbia Univ. Press, New York.

Bates, M. W., Mayer, J., and Mauss, S. F. (1955) Fat metabolism in three forms of experimental obesity. Fatty acid turnover. *Am. J. Physiol.* **180**: 309-312.

Beck, L. V., and Liu, S. C. C. (1962) Glutathione potentiation of epinephrine hyperglycemia. *Federation Proc.* **21**: 192.

Beher, W. T., and Baker, G. D. (1959) Build-up and regression of inhibitory effects of cholic acid on *in vivo* liver cholesterol synthesis. *Proc. Soc. Exptl. Biol. Med.* **101**: 214-217.

Beher, W. T., Baker, G. D., and Anthony, W. L. (1962) Feedback control of cholesterol biosynthesis in the mouse. *Proc. Soc. Exptl. Biol. Med.* **109**: 863-868.

Benditt, E. P. (1961) On the role of tanning in the pathogenesis of abnormal protein deposits: Ceruloplasmin and catecholamines as a potential tanning system. *Lab. Invest.* **10**: 1031-1039.

Benedict, S. R. (1916) Uric acid in its relations to metabolism. *Harvey Lectures, Ser.* **11**, 346-365.

Bennet, E. L., Kreech, D., Rosenzweig, M. R., Karlsson, H., Dyne, N., and Ohlander, A. (1958a) Cholinesterase and lactic dehydrogenase activity in the rat brain. *J. Neurochem.* **3**: 153-160.

Bennet, E. L., Rosenzweig, M. R., Kreech, D., Karlsson, H., Dyne, N., and Ohlander, A. (1958b) Individual, strain and age differences in cholinesterase activity of the rat brain. *J. Neurochem.* **3**: 144-152.

Bennet, E. L., Crossland, J., Kreech, D., and Rosenzweig, M. R. (1960) Strain differences in acetylcholine concentration in the brain of the rat. *Nature* **187**: 787-788.

Bennett, D. (1956) Developmental analysis of a mutation with pleiotropic effects in the mouse. *J. Morphol.* **98**: 199-233.

Bennett, D. (1959) Brain hernia, a new recessive mutation of the mouse. *J. Heredity* **50**: 265-268.

Bennett, D. (1961) A chromatographic study of abnormal urinary amino acid excretion in mutant mice. *Ann. Human Genet. (London)* **25**: 1-6.

Benson, G. K., Cowie, A. T., and Tindal, J. S. (1958) The pituitary and the maintenance of milk secretion. *Proc. Roy. Soc. (London) Ser. B* **149**: 330-336.

Bernick, S., and Patek, P. R. (1961) The effect of cholesterol feeding on the morphology of selected endocrine glands. *Arch. Pathol.* **72**: 321-330.

Bernick, S., Patek, P. R., Ershoff, B. H., and Wells, A. (1962) Effects of cholesterol feeding on the thyroid gland and vascular structures of the rabbit, guinea pig, hamster and rat. *Am. J. Pathol.* **41**: 661-670.

References

Bernstein, S. E., Russell, E. S., and Lawson, F. A. (1959) Limitations of wholebody irradiation for inducing acceptance of homografts in cases involving a genetic defect. *Transplant. Bull.* **24**: 106-108.

Beyer, R. E. (1955) A study of insulin metabolism in an insulin tolerant strain of mice. *Acta Endocrinol.* **19**: 309-332.

Biancifiori, C., and Ribacchi, R. (1962) Pulmonary tumors in mice induced by oral isoniazid and its metabolites. *Nature* **194**: 588-589.

Bidwell, E. (1955a) The purification of bovine anti-haemophilic globulin. *Brit. J. Haematol.* **1**: 35-45.

Bidwell, E. (1955b) The purification of anti-haemophilic globulin from animal plasma. *Brit. J. Haematol.* **1**: 386-389.

Bielschowsky, M., and Bielschowsky, F. (1956) New Zealand strain of obese mice; their response to stilbestrol and to insulin. *Australian J. Exptl. Biol. Med. Sci.* **34**: 181-198.

Bielschowsky, M., and Bielschowsky, F. (1962) Reaction of reticular tissue of mice with autoimmune hemolytic anemia to 2-aminofluorene. *Nature* **194**: 692-693.

Bielschowsky, M., Helyer, B. J., and Howie, J. B. (1959) Spontaneous hemolytic anemia in mice of the NZB/B1 strain. *Proc. Univ. Otago Med. School* **37**(2): 9-11.

Biggers, J. D., McLaren, A., and Michie, D. (1960) Choice of animals for bio-assay. *Nature* **190**: 891-894.

Biggs, R., Macfarlane, R. G., and Pilling, J. (1947) Observations on fibrinolysis. Increase in activity produced by exercise or adrenaline. *Lancet* **I**: 402-405.

Bigland, C. H., and Triantaphyllopoulos, D. C. (1960) A re-evaluation of the clotting time of chicken blood. *Nature* **186**: 644.

Blaschko, H. (1960) New observations on the amine oxidases of mammalian plasma. *Farmaco (Pavia) Ed. Sci.* **15**: 532.

Bleisch, V. R., Mayer, J., and Dickie, M. M. (1952) Familial diabetes mellitus in mice associated with insulin resistance, obesity and hyperplasia of the island of Langerhans. *Am. J. Pathol.* **28**: 369-385.

Blount, R. F., and Blount, I. H. (1961) Strain differences and relation of aging on salt susceptibility in mice. *Texas Rept. Biol. Med.* **19**: 739-748.

Bogart, R., and Muhrer, M. E. (1942) The inheritance of a hemophilia-like condition in swine. *J. Heredity* **33**: 59-64.

Bosanquet, F. D., Daniel, P. M., and Parry, H. B. (1956) Myopathy in sheep; its relationship to scrapie and to dermatomyositis and muscular dystrophy. *Lancet* **II**: 737-746.

Bowman, B. H., and King, F. J. (1961) Effects of glutamine and asparagine supplements in the dietary regimen of three phenylketonuric patients. *Nature* **190**: 417-418.

Boxer, G. E., and Devlin, T. M. (1961) Pathways of intracellular hydrogen transport. *Science* **134**: 1495-1501.

Boyland, E. (1958) The biological examination of carcinogenic substances. *Brit. Med. Bull.* **14**: 93-98.

Brobeck, J. R., Tepperman, J., and Long, C. N. H. (1943) Experimental hypothalamic hyperphagia in the albino rat. *Yale J. Biol. Med.* **15**: 831-853.

Brodie, B. B., Spector, S., and Shore, P. A. (1959) Interaction of monoamine oxidase inhibitors with physiological and biochemical mechanisms in brain. *Ann. N. Y. Acad. Sci.* **80**: 609-616.

Brown, A. M. (1961) Differential characteristics in uniform strains of mice. *Nature* **195**: 204-205.

Brown-Grant, K. (1961) Enlargement of salivary gland in mice treated with isopropyl-noradrenaline. *Nature* **191**: 1076-1078.
Bruell, J. H., Caroczy, A. F., and Hellerstein, H. K. (1962) Strain and sex differences in serum cholesterol levels of mice. *Science* **135**: 1071-1072.
Burian, H. M., von Noorden, G. K., and Ponsetti, I. V. (1960) Chamber angle anomalies in systemic connective tissue disorders. *A.M.A. Arch. Ophthalmol.* **64**: 671-680.
Burnet, F. M., and Holmes, M. C. (1962) Thymus lesions in an auto-immune disease of mice. *Nature* **194**: 146-147.
Burnet, F. M. (1962) Autoimmune disease—experimental and clinical. *Proc. Roy. Soc. Med.* **55**: 619-626.
Burns, M. (1952) "Genetics of the Dog." Commonwealth Agr. Bur., Edinburgh, Scotland.
Burnstein, S., Dorfman, R. I., and Nadel, E. M. (1955) Corticosteroids in the urine of normal and scorbutic guinea pigs: isolation and quantitative determination. *J. Biol. Chem.* **213**: 597-608.
Butterworth, K. R., and Mann, M. (1957) A quantitative comparison of the sympathomimetic amine content of the left and right adrenal glands of the cat. *J. Physiol. (London)* **136**: 294-299.
Butterworth, K. R., and Mann, M. (1960a) The percentage of noradrenaline in the adrenal glands of families of cats. *J. Physiol. (London)* **154**: 43P-44P.
Butterworth, K. R., and Mann, M. (1960b) Proportion of noradrenaline to adrenaline in the adrenal glands of litter-mate cats. *Nature* **187**: 785.
Calkins, E., Kahn, D., and Diner, W. C. (1956) Idiopathic familial osteoporosis in dogs: "osteogenesis imperfecta." *Ann. N. Y. Acad. Sci.* **64**: 410-423.
Canal, N., and Frattola, L. (1962) Studies on the "pentose-phosphate pathway" in hereditary muscular dystrophy in mice. *Med. Exptl.* **7**: 27-31.
Carbone, J. V., and Grodsky, G. M. (1957) Constitutional non-hemolytic hyperbilirubinemia in the rat: defect of bilirubin conjugation. *Proc. Soc. Exptl. Biol. Med.* **94**: 461-463.
Carroll, F. D., and Gregory, P. W. (1962) Responses of the Snell dwarf mouse to pituitary tissue from a bovine dwarf mutant. *Proc. Soc. Exptl. Med. Biol.* **109**: 35-38.
Carroll, F. D., Gregory, P. W., and Rollins, W. C. (1951) Thyrotropine hormone deficiency in homozygous dwarf beef cattle. *J. Animal Sci.* **10**: 916-921.
Carstensen, H., Hellman, B., and Larsson, S. (1961) Biosynthesis of steroids in the adrenals of normal and obese-hyperglycemic mice. *Acta Soc. Med. Upsalien.* **64**: 139-151.
Caspari, E. (1960) Paper presented at Am. Psychol. Assoc. Ann. Convention.
Castle, W. E. (1940) "Mammalian Genetics." Harvard Univ. Press, Cambridge, Massachusetts.
Castle, W. E. (1947) The domestication of the rat. *Proc. Natl. Acad. Sci. U.S.* **33**: 109-117.
Chai, C. K. (1957a) Leukopenia: an inherited character in mice. *Science* **126**: 125.
Chai, C. K. (1957b) Developmental homeostasis of body growth in mice. *Naturalist* **91**: 49-55.
Chai, C. K. (1957c) Analysis of quantitative inheritance of body size in mice. III. Dominance. *Genetics* **42**: 602-607.
Chai, C. K. (1958) Endocrine variation, thyroid function in inbred and F_1 hybrid mice. *J. Heredity* **44**: 143-148.
Chai, C. K. (1959) Life span in inbred and hybrid mice. *J. Heredity* **50**: 203-208.

Chai, C. K. (1960) Response of inbred and F_1-hybrid to hormone. *Nature* **185**: 514-518.
Chai, C. K. (1961) Hereditary spasticity in mice. *J. Heredity* **52**: 241-243.
Chai, C. K., and Degenhardt, K. H. (1962) Developmental anomalies in inbred rabbits. *J. Heredity* **53**: 174-182.
Chai, C. K., Amin, A., and Reineke, E. P. (1957) Thyroidal iodine metabolism in inbred and F_1 hybrid mice. *Am. J. Physiol.* **188**: 499-502.
Chai, C. K., Roberts, E., and Sidman, R. L. (1962) Influence of aminoxyacetic acid, alpha-amino-butyrate transaminase inhibitor, on hereditary spastic defects in the mouse. *Proc. Soc. Exptl. Biol. Med.* **109**: 491-495.
Chase, H. B., Gunther, M. S., Miller, J., and Wolffson, D. (1948) High insulin tolerance in an inbred strain of mice. *Science* **107**: 297-299.
Christophe, J., Dagenais, Y., and Mayer, J. (1959) Increased circulating insulin-like activity in obese-hyperglycaemic mice. *Nature* **184**: 61-62.
Christophe, J., Jeanrenaud, B., Mayer, J., and Renold, A. E. (1961a) Metabolism *in vitro* of adipose tissue in obese-hyperglycemic and goldthioglucose treated mice. I. Metabolism of glucose. *J. Biol. Chem.* **236**: 642-647.
Christophe, J., Jeanrenaud, B., Mayer, J., and Renold, A. E. (1961b) Metabolism *in vitro* of adipose tissue in obese-hyperglycemic and goldthioglucose treated mice. II. Metabolism of pyruvate and acetate. *J. Biol. Chem.* **236**: 648-652.
Cohen, M. M., Shklar, G., and Yerganian, G. (1961) Periodontal pathology in a strain of Chinese hamster, *Cricetulus griseus,* with hereditary diabetes mellitus. *Am. J. Med.* **31**: 864-867.
Coleman, D. L. (1960) Phenylalanine hydroxylase activity in dilute and non-dilute strains of mice. *Arch. Biochem. Biophys.* **91**: 300-306.
Coleman, D. L. (1961) Effects of dietary creatine and glycine on transamidinase activity in dystrophic mice. *Arch. Biochem. Biophys.* **94**: 183-186.
Coleman, D. L. (1962a) Effect of genic substitution on the incorporation of tyrosine into melanin of mouse skin. *Arch. Biochem. Biophys.* **69**: 562-568.
Coleman, D. L. (1962b) Biochemistry of muscular dystrophy. *Ann. Rept., Roscoe B. Jackson Memorial Lab. 1961-1962 No.* **33**: 11.
Coleman, D. L., and Ashworth, M. E. (1959) Incorporation of glycine-1-C^{14} into nucleic acids and proteins of mice with hereditary muscular dystrophy. *Am. J. Physiol.* **197**: 839-841.
Coleman, D. L., and Ashworth, M. E. (1960) Influence of diet on transamidinase activity in dystrophic mice. *Am. J. Physiol.* **199**: 927-930.
Committee on Standardized Genetic Nomenclature for Mice. (1960) Standardized nomenclature for inbred strains of mice: 2nd listing. *Cancer Res.* **20**: 145-169.
Conney, A. H., and Burns, J. J. (1961) Factors influencing drug metabolism. *Advan. Pharmacol.* **1**, 31-58.
Constantinides, P., Fortier, C., and Skeleton, F. R. (1950) The effect of glucose on the adrenal response to ACTH in hypophysectomized rats. *Endocrinology* **47**: 351-363.
Cook, R. P., and Thomson, R. D. (1951a) Cholesterol metabolism: cholesterol metabolism in the guinea pig and rabbit. *Biochem. J.* **49**: 72-77.
Cook, R. P., and Thomson, R. D. (1951b) The absorption of fat and of cholesterol in the rat, guinea pig, and rabbit. *Quart. J. Exptl. Physiol.* **36**: 61-74.
Cooper, J. R., and Melcer, I. (1961) Enzymic oxidation of tryptophan to 5-hydroxytryptophan in the biosynthesis of serotonin. *J. Pharmacol. Exptl. Therap.* **132**: 265-268.

Courrier, R., and Cologne, A. (1951) Cortisone et gestation chez la lapine. *Compt. rend. Acad. Sci.* **232**: 1164-1166.

Cranston, E. M. (1961) Anti-estrus drugs on subestrus of ovariectomized C3H mice. *Proc. Soc. Exptl. Biol. Med.* **108**: 514-518.

Crenshaw, W. W., Pipes, G. W., Ruppert, H. L., Jr., and Turner, C. W. (1957) Indications of normal pituitary and thyroid function in dwarf beef animals. *Missouri Agr. Expt. Sta. Res. Bull.* **621**: 1-24.

Curry, D. M., and Beaton, G. H. (1958) Cortisone resistance in pregnant rats. *Endocrinology* **63**: 155-161.

Dagg, C. P. (1960) Sensitive stages for the production of developmental abnormalities in mice with 5-fluorouracil. *Am. J. Anat.* **106**: 89-96.

Davis, B. K. (1962) The influence of alloxan diabetes, methylthiouracil, cortisone and adrenaline on the utilization of glucose-C^{14} and 1-cystine-S^{35} and mitotic activity by hair follicles in white mice. *Acta Endocrinol. Suppl.* **61**: 3-24.

Davison, A. N., and Sandler, M. (1958) Inhibition of 5-hydroxytryptophan decarboxylase by phenylalanine metabolites. *Nature* **181**: 186-187.

DeCosta, E. J., and Abelman, M. A. (1952) Cortisone and pregnancy; experimental and clinical study of effects of cortisone on gestation. *Am. J. Obstet. Gynecol.* **64**: 746-767.

Denny-Brown, D. (1960) The nature of polymyositis and related muscular diseases. *Trans. Coll. Physicians Phila.* **28**: 14-29.

Deringer, M. K., Dunn, T. B., and Heston, W. E. (1953) Results of exposure of strain C3H mice to chloroform. *Proc. Soc. Exptl. Biol. Med.* **88**: 474-479.

DeSchaepdryver, A. F., and Preziosi, P. (1959) Pharmacological depletion of adrenaline and noradrenaline in various organs of mice. *Arch. Intern. Pharmacodyn.* **111**: 177-221.

Deuel, H. J., Jr. (1955) "The Lipids," Vol. II: Biochemistry, p. 919. Wiley (Interscience), New York.

Didisheim, P., Hattori, K., and Lewis, J. H. (1959) Hematologic and coagulation studies in various animal species. *J. Lab. Clin. Med.* **53**: 866-875.

Dieke, S. H., and Richter, C. P. (1945) Acute toxicity of thiourea to rats in relation to age, diet, strain and species variation. *J. Pharmacol. Exptl. Therap.* **83**: 195-202.

Doell, R. G. (1962) Urethan induction of thymic lymphoma in C57BL mice. *Nature* **194**: 588-589.

Dole, V. P. (1961) Effects of nucleic acid metabolites on lipolysis in adipose tissue. *J. Biol. Chem.* **236**: 3125-3130.

Doolittle, R. F., and Surgenor, D. M. (1962) Blood coagulation of fish. *Am. J. Physiol.* **203**: 964-970.

Dowben, R. M., and Hsia, D. Y. Y. (1962) Inhibition of glucuronosyl transferase by steroids. *J. Lab. Clin. Med.* **60**: 870 (Abstr.).

Dowling, J. E., and Wald, G. (1958) Vitamin A deficiency and night blindness. *Proc. Natl. Acad. Sci. U.S.* **44**: 648-661.

Dowling, J. E., and Wald, G. (1960) The biological function of vitamin A acid. *Proc. Natl. Acad. Sci. U.S.* **46**: 587-608.

Elliot, F. H., and Schally, A. V. (1955) Chromatography of steroids produced by rat adrenals *in vitro*. *Can. J. Biochem. Physiol.* **33**: 174-180.

Emorson, G. (1953) Introduction to the second session. *In* "Rat Quality, A. Consideration of Heredity, Diet and Disease," pp. 102-103. The National Vitamin Foundation, New York.

Eränkö, O. (1961) Cell types of the adrenal medulla. *In* "Adrenergic Mechanisms," Ciba Foundation Symposium (G. E. W. Wolstenholme and M. O'Connor, eds.), pp. 103-108. Little, Brown, Boston, Massachusetts.

Erspamer, V. (1954) Il sistema cellulare enterocromaffine e l'enteramina (5-idrossitriptamina). *Rend. Sci. Farmitalia* **1**: 1-193.

Evans, H. M., and Simpson, M. E. (1931) Hormones of the anterior hypophysis. *Am. J. Physiol.* **98**: 511-546.

Faigle, J. W., Keberle, H., Riess, W., and Schmid, K. (1962) The metabolic fate of thalidomide. *Experientia* **18**: 389-412.

Fainstat, T. D. (1954) Cortisone-induced congenital cleft palate in rabbits. **55**: 502-508.

Falconer, D. S. (1951) Two new mutants. "Trembler and reeler" with neurological actions in the house mouse. *J. Genet.* **56**: 192-201.

Falconer, D. S., and Isaacson, J. H. (1959) Adipose, a new inherited obesity of the mouse. *J. Heredity* **50**: 290-292.

Falconer, D. S., and Isaacson, J. H. (1962) The genetics of sex-linked anemia in the mouse. *Genet. Res.* **3**: 248-250.

Fantl, P. (1961) A comparative study of blood coagulation in vertebrates. *Australian J. Exptl. Biol.* **39**: 403-412.

Fellman, J. H. (1956) Inhibition of DOPA decarboxylase by aromatic acids associated with phenylpyruvia. *Proc. Soc. Exptl. Biol. Med.* **93**: 413-414.

Fenton, P. F., and Duguid, T. R. (1962) Growth hormone and cardiac glycogen: Influence of environmental and genetic factors. *Can. J. Biochem. Physiol.* **40**: 337-341.

Field, R. A., Rickard, C. G., and Hutt, F. B. (1946) Hemophilia in a family of dogs. *Cornell Vet.* **36**: 285-300.

Figueroa, R. B., and Klotz, A. P. (1962) Alterations of alcohol dehydrogenase and other hepatic enzymes following oral alcohol intoxication. *Am. J. Clin. Nutr.* **11**: 235-239.

Fishman, W. H. (1961) "Chemistry of Drug Metabolism," pp. 31, 56, 76, 87, 128 and 130. Thomas, Springfield, Illinois.

Folch, J., Casals, J., Pope, A., Meath, J. A., LeBaron, F. N., and Les, M. (1959) Chemistry of myelin development. *In* "The Biology of Myelin" (S. R. Korey, ed.), pp. 122-137. Harper (Hoeber), New York.

Foley, C. W., Heidenreich, C. J., and Lasley, J. F. (1960) Influence of the dwarf gene on insulin sensitivity. *J. Heredity* **51**: 278-283.

Fortier, C. (1959) Relative production of fluorescent and UV-absorbing steroids by incubated rat adrenal glands. *Can. J. Biochem. Physiol.* **37**: 571-574.

Fouts, J. R. (1962) Interaction of drugs and hepatic microsomes. *Federation Proc.* **21**: 1107-1111.

Fouts, J. R., and Adamson, R. H. (1959) Drug metabolism in the newborn rabbit. *Science* **129**: 897-898.

Francis, T. (1944) Studies on hereditary dwarfism in mice. VI. Anatomy, histology and development of the pituitary at hereditary anterior pituitary dwarfism in mice. *Acta Pathol. Microbiol. Scand.* **21**: 928-944.

Francis, T. (1945) Studies in hereditary dwarfism in mice. VIII. The histology of the anterior pituitary of mice with hereditary adiposity and of dwarf mice with hereditary adiposity. *Acta Pathol. Microbiol. Scand.* **22**: 138-143.

Frankel, H. H., Patek, P. R., and Bernick, S. (1962) Long term studies of the rat reticulo-endothelial system and endocrine gland responses to foreign particles. *Anat. Record* **142**: 359-373.

Fraser, F. C., and Fainstat, T. D. (1951) Production of congenital defects in the offspring of pregnant mice treated with cortisone. *Pediatrics* **8**: 527-533.

Fredrickson, D. S., Altrocchi, P. H., Avioli, L. V., Goodman, D. S., and Goodman, H. C. (1961) Tangier disease. *Ann. Internal Med.* **55**: 1016-1031.

Fuller, J. L., and Ginsburg, B. E. (1954) Effect of adrenalectomy on the anticonvulsant action of glutomic acid in mice. *Am. J. Physiol.* **176**: 367-370.

Furth, J., Burnett, W. T., and Gadsden, E. L. (1953) Quantitative relationship between thyroid function and growth of pituitary tumors secreting TSH. *Cancer Res.* **13**: 298-307.

Garattini, S. (1961) Enhanced toxicity of serotonin in adrenalectomized rats. *Newsletter No.* **4**: 3. Institute of Pharmacology, University of Milan, Italy.

Garattini, S., Paoletti, P., and Paoletti, R. (1959) Lipid biosynthesis *in vivo* from acetate-1-C^{14} and 2-C^{14} and mevolonic-2-C^{14} acid. *Arch. Biochem. Biophys.* **84**: 253-555.

Garattini, S., Bizzi, L., Grossi, E., Paoletti, R., Poggi, M., and Vertua, R. (1961a) Pharmakologische Untersuchungen ueber die Biosynthese des Cholesterins im Tierversuch. *Fette, Seifen, Anstrichmittel* **63**: 1027-1035.

Garattini, S., Gaiardoni, P., Mostari, A., and Palma, V. (1961b) Increased toxicity of serotonin in adrenalectomized animals. *Nature* **190**: 540-541.

Garattini, S., Lamesta, L., Mostari, A., Palma, V., and Valzelli, L. (1961c) Pharmacological and biochemical effects of 5-hydroxytryptamine in adrenalectomized rats. *J. Pharm. and Pharmacol.* **13**: 385-388.

Geratz, J. D., and Graham, J. B. (1960) Plasma thromboplastic component (Christmas factor, factor IX) levels in stored human blood and plasma. *Thromb. Diath. Haemorrhag.* **4**: 376-388.

Geri, G. (1954) Considerazioni et Reserche sull'eredita dell' idroftalmia nel coniglio. *Rec. Sci.* **24**: 2299-2315.

Geri, G. (1955) Reserche sull'eredita dell' idroftalmia nel coniglio. *Riv. Zootec.* **28**: 37-42.

Gillman, T., and Gilbert, C. (1956) Periarteritis and other forms of necrotising angeitis produced by vitamin D in thyroxinised rats with an assessment of the aetiology of those vascular lesions. *Brit. J. Exptl. Pathol.* **37**: 584-596.

Gillman, T., and Hathorn, M. (1959) Sex incidence of vascular lesions in aging rats in relation to previous pregnancies. *Nature* **183**: 1139-1140.

Gillman, T., and Naidoo, S. S. (1958) Gonadal influences on plasma fibrin and fibrinolytic activity: A possible basis for further analysis of some forms of coronary thrombosis. *Endocrinology* **62**: 92-97.

Ginsburg, B. E., and Roberts, E. (1951) Glutamic acid and central nervous system activity. *Anat. Record* **111**: 76-77.

Ginsburg, B. E., Ross, S., Zamis, M. J., and Perkins, A. (1950) An assay method for the behavioral effects of *l*-glutamic acid. *Science* **112**: 12-13.

Ginsburg, B. E., Ross, S., Zamis, J. J., and Perkins, A. (1951) Some effects of *l*(+)-glutamic acid on sound-induced seizures in mice. *J. Comp. Physiol. Psychol.* **44**: 134-141.

Girard, J., Vest, M., and Roth, N. (1961) Growth hormone content of serum in infants, children, adults and hypopituitary dwarfs. *Nature* **192**: 1051-1053.
Glick, D. (1940) Properties of tropine esterase. *J. Biol. Chem.* **134**: 617-625.
Glick, D., and Glaubach, S. (1941) The occurrence and distribution of atropinesterase, and the specificity of tropinesterases. *J. Gen. Physiol.* **25**: 197.
Goldenberg, H., and Fishman, V. (1961) Species dependence of chlorpromazine metabolism. *Proc. Soc. Exptl. Biol. Med.* **108**: 178-182.
Gorbman, A., and Evans, H. M. (1943) Beginning of function in the thyroid of the fetal rat. *Endocrinology* **32**: 113-115.
Goswami, M. N. D., and Knox, W. E. (1961) Developmental changes of hydroxyphenylpyruvate-oxidase activity in mammalian liver. *Biochim. Biophys. Acta* **50**: 35-40.
Gould, A., and Coleman, D. L. (1961) Accumulation of acetoacetate in muscle homogenates from dystrophic mice. *Biochem. Biophys. Acta* **47**: 422-423.
Gould, A., and Coleman, D. L. (1962) Acetoacetate metabolism in muscle homogenates from normal and dystrophic mice. *Arch. Biochem. Biophys.* **92**: 408-411.
Graham, J. B., Buckwalter, J. A., Hartley, L. J., and Brinkhous, K. M. (1949) Canine hemophilia. Observations on the course, the clotting anomaly and the effect of blood transfusion. *J. Exptl. Med.* **90**: 97-111.
Green, E. L. (1962) "Handbook on Genetically Standardized JAX Mice," pp. 56 and 61-62.
Green, E. L., and Doolittle, D. P. (1963) Theoretical consequences of systems used in mammalian genetics. *In* "Methodology in Mammalian Genetics" (W. J. Burdette, ed.), Vol. 2. Holden-Day, San Francisco.
Green, M. C. (1961) Himalayan, a new allele of albino in the mouse. *Heredity* **52**: 73-75.
Green, M. N., Yerganian, G., and Meier, H. (1960) Elevated alpha-2-serum protein as a possible genetic marker in spontaneous hereditary diabetes mellitus of the Chinese hamster (*Cricetulus griseus*). *Experientia* **16**: 503-504.
Gregory, P. W., Rollins, W. C., Pattengale, P. S., and Carroll, F. D. (1951) Phenotypic expression of homozygous dwarfism in beef cattle. *J. Animal Sci.* **10**: 922-933.
Gregory, P. W., Julian, L. M., and Tyler, W. S. (1960a) Genetic relationships of some bovine achondroplastic mutants. *Records Genet. Soc. Am.* **29**: 72 (abstr.).
Gregory, P. W., Tyler, W. S., and Julian, L. M. (1960b) Evidence that the Dexter mutant is genetically related to recessive achondroplasia. *Anat. Record* **138**: 353-354.
Grewal, M. S. (1962) A sex-linked anemia in the mouse. *Genet. Res.* **3**: 238-247.
Grossi, E., Paoletti, P., and Paoletti, R. (1958) An analysis of brain cholesterol and fatty acid biosynthesis. *Arch. Intern. Physiol. Biochim.* **66**: 564-569.
Grüneberg, H. (1939) Inherited macrocytic anemias in the house mouse. *Genetics* **24**: 777-810.
Grüneberg, H. (1942a) The anemia of flexed tailed mice (*Mus musculus* L.) I. Static and dynamic haematology. *J. Genet.* **43**: 45-68.
Grüneberg, H. (1942b) The anemia of flexed tailed mice (*Mus musculus* L.) II. Siderocytes. *J. Genet.* **44**: 246-271.
Grüneberg, H. (1952a) "The Genetics of the Mouse." Nijhoff, The Hague.
Grüneberg, H. (1952b) Inherited macrocytic anemias in the house mouse. II. Dominance relationships. *J. Genet.* **43**: 285-293.
Grunnet, J. (1942) Studies on hereditary dwarfism in mice; investigations of tissue in islets of Langerhans in normal mice and in dwarfs. *Acta Pathol. Microbiol. Scand.* **19**: 563.

Gunn, C. H. (1938) Hereditary acholuric jaundice in a mutant strain of rats. *J. Heredity* **29**: 137-139.
Halick, P., and Seegers, W. H. (1956) Antigenic properties of purified bovine prothrombin. *Am. J. Physiol.* **187**: 103-105.
Hamburgh, M. (1958) The distribution of acetylcholinesterase in the mouse brain. *Anat. Record* **130**: 311 (abstr.).
Hanna, B. L., Sawin, P. B., and Sheppard, L. B. (1962) Recessive buphthalamos in the rabbit. *Genetics* **47**: 519-529.
Harman, D. (1956a) Reducing agents as chemotherapeutic agents in cancer. *Clin. Res.* **4**: 54-55.
Harman, D. (1956b) A theory based on free radical and radiation chemistry. *J. Gerontol.* **11**: 298-300.
Harman, D. (1957) Prolongation of the normal life span by radiation protection chemicals. *J. Gerontol.* **12**: 257-263.
Harman, D. (1961) Prolongation of the normal life span and inhibition of spontaneous cancer by antioxidants. *J. Gerontol.* **16**: 247-254.
Harman, P. J. (1950) Polycystic alterations in the white matter of the "jittery" mouse. *Anat. Record* **106**: 304 (abstr.).
Harned, B. K., and Cole, V. V. (1939) Evidence of hyperfunction of the anterior pituitary in a strain of rats. *Endocrinology* **25**: 689-697.
Harris, H. (1959) "Human Biochemical Genetics," p. 84. Cambridge Univ. Press, London and New York.
Haynes, R. C. (1958) The activation of adrenal phosphorylase by the adrenocorticotropic hormone. *J. Biol. Chem.* **233**: 1220-1222.
Haynes, R. C., and Berthet, L. (1957) Studies on the mechanism of action of adrenocorticotropic hormone. *J. Biol. Chem.* **225**: 115-124.
Haynes, R. C., Koritz, S. B., and Peron, F. G. (1959) Influence of adenosine 3',5'-monophosphate on corticoid production by rat adrenal glands. *J. Biol. Chem.* **234**: 1421-1423.
Hellman, B., Larsson, S., and Westman, S. (1961) Aspects of the glucose and aminoacid metabolism in the liver and diaphragm of normal and obese-hyperglycemic mice. *Acta Physiol. Scand.* **53**: 330-338.
Hellman, B., Larsson, S., and Westman, S. (1962) Acetate metabolism in isolated epididymal adipose tissue from obese-hyperglycemic mice of different ages. *Acta Physiol. Scand.* **56**: 189-198.
Hess, S. M., Connamacher, R. H., and Udenfriend, S. (1961) Effect of alpha-methylamine acids on catecholamines and serotonin. *Federation Proc.* **20**: 236a.
Heston, W. E. (1938) Bent-nose in the Norway rat. *J. Heredity* **29**: 437-448.
Heston, W. E. (1942a) Genetic analysis of susceptibility to induced pulmonary tumors in mice. *J. Natl. Cancer Inst.* **3**: 69-78.
Heston, W. E. (1942b) Inheritance of susceptibility to spontaneous pulmonary tumors in mice. *J. Natl. Cancer Inst.* **3**: 79-82.
Heston, W. E. (1949) Development of inbred strains in the mouse and their use in cancer research. *In* "Lectures on Genetics, Cancer, Growth and Social Behavior," pp. 9-13. Roscoe B. Jackson Memorial Laboratory, Twentieth Commemoration, Bar Harbor, Maine.
Heston, W. E. (1950) Carcinogenic action of the mustards. *J. Natl. Cancer Inst.* **11**: 415-423.

Heston, W. E. (1956) Effects of genes located in chromosomes III, V, VII, IX and XIV on the occurrence of pulmonary tumors in the mouse. *Cytologia Suppl.* **1956**: 219-224.

Heston, W. E., and Vlahakis, G. (1961) Influence of the A^y gene on mammary tumors, hepatomas and normal growth in mice. *J. Natl. Cancer Inst.* **26**: 969-983.

Hillarp, N. A. (1961) Some problems concerning the storage of catecholamines in the adrenal medulla. *In* "Adrenergic Mechanisms," Ciba Foundation Symposium (G. E. W. Wolstenholme and M. O'Connor, eds.), pp. 481-486. Little, Brown, Boston, Massachusetts.

Hirsch, C. W., Kuntzman, R., and Costa, E. (1962) Effects of DOPA-5HTP decarboxylase inhibition on synthesis of brain amines. *Federation Proc.* **21**: 364.

Hobaek, A. (1961) "Problems of Hereditary Chondroplasias." Oslo Univ. Press.

Hoffman, R. A. (1958) Temperature response of the rat to action and interaction of chlorpromazine, reserpine and serotonia. *Am. J. Physiol.* **195**: 755-758.

Hofman, F. G., and Christie, N. P. (1961) Studies *in vitro* of steroid hormones biosynthesis in adrenal cortical tumors of mice. *Biochim. Biophys. Acta* **54**: 354-356.

Hogan, A. G., Muhrer, M. E., and Bogart, R. (1941) A hemophilia-like disease in swine. *Proc. Soc. Exptl. Biol. Med.* **48**: 217-219.

Hollifield, G., Perlman, M., and Parson, W. (1962) Free fatty acid content of adipose tissue in three types of obese mice during fasting. *Metabolism* **11**: 117-122.

Holmes, M. C., Gorrie, J., and Burnet, F. M. (1961) Transmission by splenic cells of an autoimmune disease occurring spontaneously in mice. *Lancet* **ii**: 638-639.

Homburger, F., Baker, J. R., Nixon, C. W., and Whitney, R. (1962) Primary, generalized polymyopathy and cardiac necrosis in an inbred line of Syrian hamster. *Med. Exptl.* **6**: 339-345.

Homburger, F., Grossman, M. S., and Tregier, A. (1955) Experimental hydrouteri (hydrometra) in rodents and some factors determining their formation. *Proc. Soc. Exptl. Biol. Med.* **90**: 719-723.

Homburger, F., and Tregier, A. (1957) Endocrine factors determining the rate of accumulation of endometrial secretions in experimental hydrouteri of mice. *Endocrinology* **61**: 627-633.

Homburger, F., Tregier, A., and Grossman, M. S. (1957) Biologic activities of endometrial secretions (uterone) collected from experimental hydrouteri of mice. *Endocrinology* **61**: 634-642.

Homburger, F., Tregier, A., Baker, J. R., and Crooker, C. M. (1961). The use of hairless mice for study of cosmetics. *Proc. Sci. Sect. Toilet Goods Assoc.* **35** (May).

Hrubant, H. E. (1959) A chromatographic analysis of the free amino acids in the blood plasma of three inbred strains of the house mouse, *Mus musculas*. *Genetics* **44**: 591-608.

Huber, H. L. (1918) The pharmacology and toxicology of copper salts of amino acids. *J. Pharmacol. Exptl. Therap.* **11**: 303-329.

Huff, S. D. (1962) A genetically controlled response to the drug chlorpromazine. *Roscoe B. Jackson Memorial Lab. Ann. Rept. 1961-1962, No.* **33**.

Huff, S. D., and Huff, R. L. (1962) The dilute locus and andiogenic seizures in mice. *Science* **136**: 318-319.

Hummel, K. P. (1961) Diabetes insipidus. *Roscoe B. Jackson Memorial Lab. Ann. Rept. 1960-1961, No.* **32**: 39-40.

Hummel, K. P., and Chapman, D. B. (1960-1961) Lesions in the pituitary gland. *Roscoe B. Jackson Memorial Lab. Ann. Rept. 1960-1961, No.* **32**: 40-41.

Hummel, K. P. (1958) Accessory adrenal cortical nodules in the mouse. *Anat. Record* **132**: 281-291.

Hummel, K. P. (1962) Heart lesions. *Roscoe B. Jackson Memorial Lab. Ann. Rept. 1961-1962, No.* **33**: 31.

Hutt, F. B. (1949) "Genetics of the Fowl," p. 16. McGraw-Hill, New York.

Hutt, F. B., Rickard, C. G., and Field, R. A. (1948) Sex-linked hemophilia in dogs. *J. Heredity* **39**: 3-9.

Ingalls, A. M., Dickie, M. M., and Snell, G. D. (1950) Obese, a new mutation in the house mouse. *J. Heredity* **41**: 317-318.

Inscoe, J. K., and Axelrod, J. (1960) Some factors affecting glucuronide formation in vitro. *J. Pharmacol. Exptl. Therap.* **129**: 128-131.

Irwin, S. (1962) Drug screening and evaluative procedures. *Science* **136**: 123-128.

Israels, M. C. G., Lempert, H., and Gilbertson, E. (1951) Hemophilia in the female. *Lancet* i: 1375.

Jacquez, J. A., and Sherman, J. H. (1962) Enzymatic degradation of azaserine. *Cancer Res.* **22**: 56-61.

Jara, Z. (1957) The blood coagulation system in carp (*Cyprinus carpio* L.). *Zool. Polon.* **8**: 113-129.

Jay, G. E., Jr. (1953) The use of inbred strains in biological research. *In* "Rat Quality, A Consideration of Heredity, Diet and Disease," pp. 98-101. The National Vitamin Foundation, New York.

Jay, G. E., Jr. (1955) Variation in response to various mouse strains to hexobarbital (Evipal). *Proc. Soc. Exptl. Biol. Med.* **90**: 378-380.

Johannesson, T., and Lausen, H. H. (1961) Chlorpromazine as an inhibitor of brain cholinesterases. *Acta Pharmacol. Toxicol.* **18**: 398-406.

Johnson, L., Blanc, W. A., Lucey, J. F., and Day, R. (1957) Kernicterus in rats with familial jaundice. *Am. J. Diseases Children* **94**: 548-556.

Johnson, L., Sarmiento, F., Blanc, W. A., and Day, R. (1959) Kernicterus in rats with an inherited deficiency of glucuronyl transferase. *Am. J. Diseases Children* **97**: 591-608.

Johnson, L. E., Harshfield, G. S., and McCone, W. (1950) Dwarfism, an hereditary defect in beef cattle. *J. Heredity* **41**: 177-181.

Julian, L. M., and Asmundson, V. S. (1960) Anatomical expression of genes of muscular dystrophy in heterozygous chickens. *Anat. Record* **136**: 218.

Julian, L. M., Tyler, W. S., Hage, T. S., and Gregory, P. W. (1957) Premature closure of the sphenooccipital synchrondrosis in the horned Hereford dwarf of the "short headed" variety. *Am. J. Anat.* **100**: 269-287.

Kalow, W. (1962) "Pharmacogenetics, Heredity and the Response to Drugs." Saunders, Philadelphia, Pennsylvania.

Kandutsch, A. A. (1962a) Enzymatic reduction of the delta-7 bond of 7-dihydrocholesterol. *J. Biol. Chem.* **237**: 358-362.

Kandutsch, A. A. (1962b) Steroid metabolism in tumors and normal tissues. *Roscoe B. Jackson Memorial Lab. Ann. Rept., 1961-1962, No.* **33**: 6-7.

Kandutsch, A. A., and Russell, A. E. (1959) Preputial gland tumor sterols. I. The occurrence of 24, 25-dihydro-lanosterol and a comparison with liver and the normal gland. *J. Biol. Chem.* **234**: 2037-2042.

References

Kandutsch, A. A., and Russell, A. E. (1960a) Preputial gland tumor sterols. II. The identification of 4-α-methyl-Δ^8-cholesten-3 β-01. *J. Biol. Chem.* **235**: 2253-2255.

Kandutsch, A. A., and Russell, A. E. (1960b) Preputial gland tumor sterols. III. A metabolic pathway from lanosterol to cholesterol. *J. Biol. Chem.* **235**: 2256-2261.

Kaplan, N. O., Diotti, M. M., Hamolsky, M., and Bieber, R. E. (1960) Molecular heterogeneity and evaluation of enzymes. *Science* **131**: 392-397.

Karki, N. T., Kuntzman, R., and Brodie, B. B. (1960) Norepinephrine and serotonin brain levels at various stages of ontogenetic development. *Federation Proc.* **19**: 282.

Kato, R., Chiesara, E., and Vassanelli, P. (1961) Metabolic differences of carisoprodol in the rat in relation to sex. *Med. Exptl.* **4**: 387-392.

Kato, R., Chiesara, E., and Frontino, G. (1962a) Influence of sex difference on the pharmacologic action and metabolism of some drugs. *Biochem. Pharmacol.* **11**: 221-227.

Kato, R., Vassanelli, P., and Frontino, G. (1962b) Activities of some hepatic microsomal TPNH-dependent steroid-metabolizing enzymes of rats in relation to age. *Excerpta Med.* **51**: 155 (Abstr.).

Katz, L. N., and Stamler, J. (1953) "Experimental Atherosclerosis," pp. 258-261. Thomas, Springfield, Illinois.

Kehl, R., Audibert, A., Gage, C., and Amarger, J. (1957) Action de la réserpine à différentes périodes de la gestation, chez la lapine. *Compt. Rend. Soc. Biol.* **159**: 2196-2199.

Kelleher, P. C., and Villee, C. A. (1962) Serum-protein changes in the foetal rat studied by immunoelectrophoresis. *Biochim. Biophys. Acta* **59**: 252-254.

Kelton, E. D. (1961) A study of lethal alleles of dilute in the house mouse. Ph.D. Thesis, University of Massachusetts, Amherst, Massachusetts.

King, L. S. (1936) Hereditary defects of the corpus callosum of the mouse, *Mus musculus. J. Comp. Neurol.* **64**: 337.

King, L. S., and Keeler, C. E. (1932) Absence of corpus callosum, A hereditary brain anomaly of the house mouse. Preliminary Report. *Proc. Natl. Acad. Sci. U.S.* **18**: 525.

Kirkman, H., and Bacon, R. L. (1950) Malignant renal tumors in male hamsters (*Cricetus auratus*) treated with estrogen. *Cancer Res.* **10**: 122-123.

Knox, W. E. (1962) Enzymes in young animals. *In* "Enzymes and Drug Action," Ciba Foundation Symposium (J. L. Mongar and A. V. S. de Reuck, eds.), p. 509. Little, Brown, Boston, Massachusetts.

Koneff, A. A., Nichols, C. W., Jr., Wolff, J., and Chaikoff, I. L. (1949) The fetal bovine thyroid: morphogenesis as related to iodine accumulation. *Endocrinology* **45**: 242-249.

Koneff, A. A., Simpson, M. E., and Evans, H. M. (1946) Effects of chronic administration of diethyl stilbestrol on the pituitary and other endocrine organs of hamsters. *Anat. Record* **94**: 169-187.

Kuntzman, R., Costa, E., Creveling, C. R., Hirsch, C., and Brodie, B. B. (1962) Blockade *in vivo* of dopamine beta hydroxylase by NSD1024 and NSD1034, isosteres of alpha-methyl-meta-tyramine. *Federation Proc.* **21**: 365.

Kunze, H. (1954) Die Erythropoese bei einer erblichen Anaemie rontgenmutierter Mause. *Folia Haematol.* **72**: 392-436.

Kupperman, H. S., and Greenblatt, R. B. (1947) Relationship of sex steroids of the adrenal glands of hampsters and rats. *Endocrinology* **40**: 452.

Kyle, W. K., and Chapman, A. B. (1953) Experimental check of the effectiveness of selection for a quantitative character. *Genetics* **38**: 421-443.

Lacassagne, A. (1940) Réactions de la glande sous-maxillaire à l'hormone mâle, chez la souris et le rat. *Compt. Rend. Soc. Biol.* **133**: 539-540.

Laki, K., Kominz, D. R., Symonds, P., Lorand, L., and Seegers, W. H. (1954) The amino acid composition of bovine prothrombin. *Arch. Biochem. Biophys.* **49**: 276-282.

Lambert, W. V., and Sciuchetti, A. M. (1935) A dwarf mutation in the rat. A Mendelian recessive character possibly due to a defect of one of the endocrine glands. *J. Heredity* **26**(2): 91-94.

Landauer, W. (1932) Studies on the creeper fowl. III. The early development and lethal expression of homozygous creeper embryos. *J. Genet.* **25**: 367-394.

Lane, P. W. (1961) Lists of mutant genes and mutant-bearing stocks of the mouse. List 3: Mutant genes on inbred backgrounds. Mimeographed pamphlet, Roscoe B. Jackson Memorial Laboratory, Bar Harbor, Maine.

Lane, P. W. (1961) *Mouse News Letter* **24**: 35.

Lane, P. W., and Dickie, M. M. (1958) The effect of restricted food intake on the life span of genetically obese mice. *J. Nutr.* **64**: 549-554.

Laney, F., and Waugh, D. F. (1953) Certain physical properties of bovine prothrombin. *J. Biol. Chem.* **203**: 489-499.

Larsson, S., Hellman, B., and Carstensen, H. (1962) *In vitro* utilization of uniformly labelled C^{14}-glucose in the adrenals of normal and obese hyperglycemic mice. *Acta Endocrinol.* **39**: 599-604.

Law, L. W., Morrow, A. G., and Greenspan, E. M. (1952) Inheritance of low liver glucuronidase activity in the mouse. *J. Natl. Cancer Inst.* **12**: 909-916.

Lawe, J. E. (1962) Renal changes in hamster with hereditary diabetes mellitus. *Arch. Pathol.* **73**: 88-96.

Leboeuf, B., Lochaya, S., Leboeuf, N., Wood, F. C., Jr., Mayer, J., and Cahill, G. F. (1961) Glucose metabolism and mobilization of fatty acids by adipose tissue from obese mice. *Am. J. Physiol.* **201**: 19-22.

Lessin, A. W., and Parkes, M. W. (1957) The relationship between sedation and body temperature in the mouse. *Brit. J. Pharmacol.* **12**: 245-250.

Levy, J., and Michael, E. (1938) Sur l'hydrolyse enzymatique de l'atropine. *Compt. Rend. Soc. Biol.* **129**: 820-822.

Lillis, D. V., Miller, V. L., Bearse, G. E., and Hamilton, C. M. (1963) Differences in copper retention in two strains of chickens. *Toxicol. Appl. Pharmacol.* **5**: 12-15.

Llach, J. L., Tramezzani, J. H., and Cordero Funes, J. R. (1960) A sexual difference in the concentration of iodine-131 by the sub-maxillary gland of mice. *Nature* **188**: 1204-1205.

Lochaya, S., Leboeuf, N., Mayer, J., and Leboeuf, B. (1961) Adipose tissue metabolism of obese mice on standard and high-fat diets. *Am. J. Physiol.* **201**: 23-26.

Loosli, R., Russell, E. S., Silvers, W. K., and Southard, J. L. (1961) Variability of incidence and clinical manifestation of mouse hereditary muscular dystrophy on heterogeneous genetic backgrounds. *Genetics* **46**: 347-355.

Lorenz, E., Heston, W. E., Deringer, M. K., and Eschenbrenner, A. B. (1946) Increase in incidence of lung tumors in strain A mice following long continued irradiation with gamma-rays. *J. Natl. Cancer Inst.* **6**: 349-353.

Lucas, D. R. (1958) Retinal dystrophy strains. *Mouse News Letter* **19**: 43.

Luce, S. A. (1959) The fine structure of the morphogenesis of myelin. *In* "The Biology of Myelin" (S. R. Korey, ed.), pp. 58-81. Harper (Hoeber), New York.

Luecke, R. W., Palmer, L. S., and Kennedy, C. (1945) Effects of thiamine and ribo-

flavin deficient diets on rats differing in their efficiency of food utilization. *Arch. Biochem.* **5**: 395-400.
Maas, J. W. (1962) Neurochemical differences between two strains of mice. *Science* **137**: 621-622.
McClearn, G. E., and Rogers, D. A. (1959) Differences in alcohol preferences among inbred strains of mice. *Quart. J. Studies Alc.* **20**: 691-695.
Macfarlane, R. G., and Biggs, R. (1959) The treatment of hemophilic patients with animal antihaemophilic globulin. *In* "Hemophilia and Other Hemorrhagic States" (K. M. Brinkhous and P. DeNicola, eds.), pp. 19-26. The Univ. of North Carolina Press, Chapel Hill, North Carolina.
McGaughey, C. (1961) Excretion of uncrystallized urinary calculi composed of glycolipoprotein by normal and muscular dystrophic mice. *Nature* **192**: 1267-1269.
Madjerek, Z., and van der Vries, J. (1961) Carbonic anhydrase activity in the uteri of mice under various experimental conditions. *Acta Endocrinol.* **38**: 315-320.
Mandel, P., and Harth, S. (1961) Free nucleotides of the brain in various mammals. *J. Neurochem.* **8**: 116-125.
Margulis, R. R., Hodgkinson, C. P., Rattan, W. C., and Jewell, J. S. (1954) Toxemia of pregnancy-ACTH and cortisone therapy. *Am. J. Obstet. Gynecol.* **67**: 1237-1248.
Marks, B. H., Alpert, M., and Kruger, F. A. (1958) Effect of amphenone upon steroidogenesis in the adrenal cortex of the golden hamster. *Endocrinology* **63**: 75-81.
Marlowe, T. J. (1960) Comparison of the growth hormone content of the pituitary glands from dwarf and normal beef calves. *J. Animal Sci.* **19**: 810-819.
Marlowe, T. J., and Chambers, D. J. (1954) Some endocrine aspects of dwarfism in beef cattle. *J. Animal Sci.* **13**: 961 (Abstr.).
Marquardt, W. (1949) Die Klinik und Roentgenologie der angeborenen enchondralen Verknoecherungsstoerungen. *Fortschr. Roentgenstr.* **71**: 511-535.
Marshall, N. B., and Engel, F. L. (1960) The influence of epinephrine on adipose tissue content and release of fatty acids in obese-hyperglycemic and lean mice. *J. Lipid. Res.* **1**: 339-342.
Marshall, N. B., Barnett, R. J., and Mayer, J. (1955) Hypothalamic lesions in goldthioglucose injected mice. *Proc. Soc. Exptl. Biol. Med.* **90**: 240-244.
Martin, G. J., and Gardner, R. E. (1935) The trichogenic action of the sulfhydryl group in hereditary hypotrichosis of the rat. *J. Biol. Chem.* **111**: 193.
Martinez, A., and Sirlin, J. L. (1955) Neurohistology of the agitans mouse. *J. Comp. Neurol.* **103**: 131.
Mayer, J. (1955) Physiological basis of obesity and leanness. *Nutr. Abstr. Rev.* **25**: 597-611.
Mayer, J., Bates, M. W., and Dickie, M. M. (1951) Hereditary diabetes in genetically obese mice. *Science* **113**: 746-747.
Meier, H. (1961) Some physico-chemical properties of red blood cells and hemoglobin from inbred strains of mice. *Experientia* **16**: 215-217.
Meier, H. (1963) Potentialities for and present status of pharmacologic research in genetically controlled mice. *Advan. in Pharmacol.* **2**: in press.
Meier, H., and Hoag, W. G. (1962a) Pharmacologic reactions of genetically controlled mice. *Federation Proc.* **21**: 174.
Meier, H., and Hoag, W. G. (1962b) Activity of lysozyme in inbred mice. *J. Bacteriol.* **83**: 689-690.

Meier, H., and Hoag, W. G. (1962c) The neuropathology of "reeler," a neuromuscular mutation in the mouse. *J. Neuropathol. Exptl. Neurol.* **21**: 649-654.

Meier, H., and Hoag, W. G. (1962d) Studies on left auricular thrombosis in inbred mice. *Exptl. Med. Surg.* **19**: 317-322.

Meier, H., and Hoag, W. G. (1962e) Blood proteins and immune response in mice with hereditary absence of spleen. *Naturwissenschaften* **49**: 329-331.

Meier, H., and Yerganian, G. (1959) Spontaneous hereditary diabetes mellitus in the Chinese hamster (*Cricetulus griseus*). I. Pathological findings. *Proc. Soc. Exptl. Biol. Med.* **100**: 810-814.

Meier, H., and Yerganian, G. (1960a) Spontaneous hereditary diabetes mellitus in the Chinese hamster (*Cricetulus griseus*). II. Findings in the offspring of diabetic parents. *Diabetes* **10**: 12-18.

Meier, H., and Yerganian, G. (1960b) Spontaneous hereditary diabetes mellitus in the Chinese hamster (*Cricetulus griseus*). III. Maintenance of a diabetic hamster colony with the aid of hypoglycemic agents. *Diabetes* **10**: 19-21.

Meier, H., Allen, R. C., and Hoag, W. G. (1961) Normal blood clotting of inbred mice. *Am. J. Physiol.* **201**: 375-378.

Meier, H., Jordan, E., and Hoag, W. G. (1962a) The zymogram technique as a tool for study of genetic differences. *J. Histochem. Cytochem.* **10**: 103-104.

Meier, H., Allen, R. C., and Hoag, W. G. (1962b) Spontaneous hemorrhagic diathesis in inbred mice due to single or multiple "prothrombin-complex" deficiencies. *Blood* **19**: 501-514.

Meier, H., Allen, R. C., and Hoag, W. G. (1963) Separation and purification of clotting factors from inbred mice. *Clin. Chim. Acta* **8**: 137-139.

Meites, J. (1957) Le déclenchement de la lactation au moment de la parturition. *Ann. Endocrinol. (Paris)* **17**: 519-525.

Mengel, C. E., and Kelly, M. G. (1961) Biochemical and morphological changes in a mast cell neoplasm during treatment with cyclophosphamide. *Cancer Res.* **21**: 1545-1550.

Merskey, C. (1950) Hemophilia occurring in the human female. *Proc. Intern. Soc. Hematol.* **104**: 441-445.

Mickelsen, O., Takahashi, S., and Craig, C. (1955) Experimental obesity. I. Production of obesity in rats by feeding high-fat diets. *J. Nutr.* **57**: 541-554.

Michelson, A. M., Russell, E. S., and Harman, P. J. (1955) Dystrophia muscularis: a hereditary primary myopathy in the house mouse. *Proc. Natl. Acad. Sci. U.S.* **41**: 1079-1084.

Miller, K. D., and Seegers, W. H. (1956) The preparation of a carbohydrate fraction from prothrombin and its chemical nature. *Arch. Biochem. Biophys.* **60**: 398-401.

Miller, V. L., Bearse, G. E., Kimura, Y., and Csonke, E. (1959) Strain differences in mercury retention by chicks. *Proc. Soc. Exptl. Biol. Med.* **100**: 301-303.

Mirand, E. A., and Osborne, C. M. (1953) Insulin sensitivity in the hereditary hypopituitary dwarf mouse. *Proc. Soc. Exptl. Biol. Med.* **82**: 746-748.

Mintz, B. (1957) Embryologic development of primordial germ-cells in the mouse: Influence of a new mutation, Wj. *J. Embryol. Exptl. Morphol.* **5**: 396-403.

Mintz, B., and Russell, E. S. (1957) Gene-induced embryological modifications of primordial germ cells in the mouse. *J. Exptl. Zool.* **134**: 207-237.

Mixter, R., and Hunt, H. R. (1933) Anemia in the flexed tailed mouse, *Mus musculus*. *Genetics* **18**: 367-387.

References

Mollenback, C. J. (1940) Studies on hereditary dwarfism in mice. IV. On the function of metabolic active hormones in the anterior pituitary dwarf mouse. *Acta Pathol. Microbiol. Scand.* **18**: 169-185.

Montagna, W., Chase, H. B., and Malaragno, H. P. (1952) The skin of hairless mice. I. The formation of cysts and the distribution of lipids. *J. Invest. Dermatol.* **19**: 83-94.

Montagna, W., Chase, H. B., and Brown, P. J. (1954) The skin of hairless mice. *J. Invest. Dermatol.* **23**: 259-269.

Mordkoff, A. M., and Fuller, J. L. (1959) Variability in activity within inbred and cross bred mice. *J. Heredity* **50**: 6-8.

Mori, K. (1961) Preliminary note on adenocarcinoma of the lung in mice induced with 4-nitroquinoline N-oxide. *Gann* **52**: 265-270.

Mori, K., and Yasuno, A. (1961) Introduction of pulmonary tumors in mice by subcutaneous injection of 4-nitroquinoline N-oxide. *Gann* **52**: 149-153.

Morris, H. P., Palmer, L. S., and Kennedy, C. (1933) An experimental study of inheritance as a factor influencing food utilization in the rat. *Univ. of Minn. Agr. Expt. Sta., Tech. Bull. No.* **92**.

Mosbach, E. H., Jackel, S. S., and King, C. G. (1950) Contrasts in ascorbic acid and glucuronic acid synthesis by albino rats of the Sherman and Wistar strains. *Arch. Biochem.* **29**: 348-353.

Motulsky, A. G. (1958) Drug reactions, enzymes and biochemical genetics. *J. Am. Med. Assoc.* **165**: 835-837.

Mustard, J. F., Rowsell, H. C., Robinson, G. A., Hoeksema, T. D., and Downie, H. G. (1960) Canine hemophilia B (Christmas disease) *Brit. J. Haematol.* **6**: 259-266.

Mustard, J. F., Secord, D., Hoeksema, T. D., Downie, H. G., and Rowsell, H. C. (1962a) Canine factor VII-deficiency. *Brit. J. Haematol.* **8**: 43-47.

Mustard, J. F., Basser, W., Hedgardt, G., Secord, D., Rowsell, H. C., and Downie, H. G. (1962b) A comparison of the effect of serum and plasma transfusions on the clotting defect in canine hemophilia B. *Brit. J. Haematol.* **8**: 36-42.

Nachmias, V. T. (1961) 5-Methyltryptophan and darkening of the hair in yellow A^y/a mice. *Nature* **191**: 934-935.

Nachtsheim, H. (1958) Problems of comparative genetics in mammals. *Proc. 10th Intern. Congr. Genet.* **1**: 187-198.

Nadel, E., Burstein, L. S., and Hoagland, H. (1957) Comparative effects of ACTH and LSD on urinary corticosteroids in guinea pigs. *Am. J. Physiol.* **189**: 73-74.

Nakahara, W., Fukuoka, F., and Sugimura, T. (1957) Carcinogenic action of 4-nitroquinoline-N-oxide. *Gann* **48**: 129-137.

Nathaniel, D. R., Bernick, S., and Patek, P. R. (1963) The effects of sex hormones upon the R-E cell response to foreign particles and related arterial disease. In preparation. Cited by Bernick et al. (1962).

Niu, M. C., Cordova, C. C., Niu, L. C., and Radbill, C. L. (1962) RNA-induced biosynthesis of specific enzymes. *Proc. Natl. Acad. Sci. U.S.* **48**: 1964-1969.

Nour-Eldin, F., and Wilkinson, J. F. (1958) Changes in the blood-clotting defect in Christmas disease after plasma and serum transfusions. *Clin. Sci.* **17**: 303-308.

O'Steen, W. K., and Hrachovy, J. J. (1962) The distribution of mast cells in mice with hereditary muscular dystrophy. *Texas Repts. Biol. Med.* **20**: 70-78.

Paigen, K. (1960) The influence of a single gene on enzymatic structure of cytoplasmic particles. *Acta Unio Intern. Contra Cancrum* **16**: 1032-1034.

Paigen, K. (1959) Genetic influences on enzyme localization. *J. Histochem. Cytochem.* **7**: 248-249.

Paoletti, R., Smith, R. L., Maickel, R. P., and Brodie, B. B. (1961) Identification and physiological role of norepinephrine in adipose tissue. *Biochem. Biophys. Res. Commun.* **5**: 424-429.

Pare, C. M. B., Sandler, M., and Stacey, R. S. (1957) 5-Hydroxytryptamine deficiency in phenylketonuria. *Lancet* **i**: 551-553.

Pearce, L., and Brown, W. H. (1945) Hereditary achondroplasia in the rabbit. *J. Exptl. Med.* **82**: 241-295.

Pickens, M., Anderson, W. E., and Smith, A. H. (1940) The composition of gains made by rats on diets promoting different rates of gain. *J. Nutr.* **20**: 351-365.

Pinninger, J. L., and Franks, R. B. (1951) Hemophilia in the female. *Lancet* **ii**: 82 Letter to the Editor.

Pitt-Rivers, R. (1960) Some factors that affect thyroid hormone synthesis. *Ann. N. Y. Acad. Sci.* **86**: 362-372.

Plotnikoff, N. P. (1958) Bioassay of potential tranquilizers and sedative agents against audiogenic seizures in mice. *Arch. Intern. Pharmacodyn.* **64**: 130-135.

Plotnikoff, N. P. (1960) Ataractics and strain differences in audiogenic seizures in mice. *Pharmacologia* **1**: 429-432.

Plotnikoff, N. (1961) Drug resistance due to inbreeding. *Science* **134**: 1881-1882.

Plotnikoff, N. P., and Green, D. M. (1957) Bioassay of potential ataraxic agents against audiogenic seizures in mice. *J. Pharmacol. Exptl. Therap.* **119**: 294-298.

Poel, W. E., and Yerganian, G. (1961) Adenocarcinoma of the pancreas in diabetes-prone Chinese hamsters. *Am. J. Med.* **31**: 861-863.

Poiley, S. M. (1953) History and information concerning the rat colonies in the Animal Section of the National Institutes of Health. *In* "Rat Quality, A Consideration of Heredity, Diet and Disease," pp. 86-87. The National Vitamin Foundation, New York.

Popp, R. A., and Popp, D. M. (1962) Inheritance of serum esterases having different electrophoretic patterns. *J. Heredity* **53**: 111-114.

Price Evans, D. A., and Clarke, C. A. (1961) Pharmacogenetics. *Brit. Med. Bull.* **17**: 234-240.

Putilin, K. I. (1929) *Zhur. Eksptl. Biol. i Med.* **5**: 1; quoted by Allison *et al.* (1957).

Quick, A. J. (1959) *In* "Hemorrhagic Diseases," p. 186. Lea & Febiger, Philadelphia, Pennsylvania.

Radev, T. (1960) Inheritance of hypocatalasemia in guinea pigs. *Genetics* **57**: 169-172.

Rankin, R. M. (1941) Changes in the content of iodine compounds and in the histological struction of the thyroid gland of the pig during fetal life. *Anat. Record* **80**: 123-125.

Reed, J. (1951) A study of the alcoholic consumption and amino acid excretion patterns of rats of different inbred strains. *Univ. Texas Publ.* **5109**: 144-149.

Reiss, M., Brummel, E., Halkerston, I. D. K., Badrick, F. E., and Fenwick, M. (1953) The *in vitro* action of ACTH on the oxygen consumption of slices of cattle adrenal cortex. *J. Endocrinol.* **9**: 379-390.

Renson, J., Goodwin, F., Weissbach, H., and Udenfriend, S. (1961) Conversion of tryptophan to 5-hydroxy-tryptophan by phenylalanine hydroxylase. *Biochem. Biophys. Res. Commun.* **6**: 20-23.

Roberts, E. (1924/1925) Inheritance of hypotrichosis in rats. *Anat. Record* **29**: 141.

Roberts, E. (1926/1927) Further data on inheritance of hypotrichosis in rats. *Anat. Record* **34**: 172.

Roberts, E., and Borges, P. (1955) Patterns of free amino acids in growing and regressing tumors. *Cancer Res.* **15**: 697-699.

Roberts, E., and Frankel, S. (1949) Free amino acids in normal and neoplastic tissues of mice as studied by paper chromatography. *Cancer Res.* **9**: 645-648.

Robinson, R. (1957) Partial albinism in the Syrian hamster. *Nature* **180**: 443-444.

Roderick, T. H. (1960) Selection for cholinesterase activity in the cerebral cortex of the rat. *Genetics* **45**: 1123-1140.

Roderick, T. H., and Storer, J. B. (1961) Correlation between mean litter size and mean life span among 12 inbred strains of mice. *Science* **134**: 48-49.

Rogers, S. (1951) Age of host and other factors affecting the production with urethan of pulmonary adenomas in mice. *J. Exptl. Med.* **93**: 427-449.

Rosen, F. (1959) The relationship of certain vitamin deficiencies to the toxicity of iproniazid. *Ann. N. Y. Acad. Sci.* **80**: 885-897.

Rosengren, E., and Steinhardt, C. (1961) Elevated histidine decarboxylase activity in the kidney of the pregnant mouse. *Experientia* **17**: 544-545.

Rosenzweig, M. R., Kreech, D., and Bennet, E. L. (1958) *In* "Neurological Basis of Behavior," Ciba Foundation Symposium (G. E. Wolstenhohne and C. M. O'Connor, eds.), p. 337. Little, Brown, Boston, Massachusetts.

Runner, M. N., and Dagg, C. P. (1959) Metabolic mechanisms of teratogenic agents during morphogenesis. *Natl. Cancer Inst. Monograph No.* **2**: 41-54.

Runner, M. N., and Gates, A. (1954) Sterile obese mothers. *J. Heredity* **45**: 51-58.

Russell, E. S. (1949) Analysis of pleiotropism at the W-locus in the mouse: relationship between the effects of W and W^v substitution on hair pigmentation and on erythrocytes. *Genetics* **34**: 708-724.

Russell, E. S. (1955) Review of the pleiotropic effects of W-series genes on growth and differentiation. *In* "Aspects of Synthesis and Order in Growth" (D. Rudnick, ed.), Chapter V, pp. 113-126. Princeton Univ. Press, Princeton, New Jersey.

Russell, E. S. (1960) New trends in the use of genetically controlled animals in biomedical research. *Proc. Animal Care Panel* **10**: 167-176.

Russell, E. S. (1961) Genetic studies of muscular dystrophy in the house mouse. *Proc. 2nd Intern. Conf. Human Genetics, Rome.*

Russell, E. S., and Fekete, E. (1958) Analysis of W-series pleiotropism in the mouse: effect of W^vW^v substitution on definitive germ cells and on ovarian tumorigenesis. *J. Natl. Cancer Inst.* **21**: 365-381.

Russell, E. S., and Fondal, E. L. (1951) Quantitative analysis of the normal and four alternative degrees of an inherited macrocytic anemia in the house mouse. *Blood* **6**: 892-905.

Russell, E. S., and Gerald, P. S. (1958) Inherited electrophoretic hemoglobin patterns among 20 inbred strains of mice. *Science* **128**: 1569-1570.

Russell, E. S., and Lawson, F. A. (1959) Selection and inbreeding for longevity of a lethal type. *J. Heredity* **50**: 19-25.

Russell, E. S., Smith, L. J., and Lawson, F. A. (1956) Implantation of normal blood-forming tissue in radiated genetically anemic hosts. *Science* **124**: 1076-1077.

Russell, E. S., Lawson, F., and Schabtach, G. (1957) Evidence for a new allele at the W-locus of the mouse. *J. Heredity* **48**: 119-123.

Russell, E. S., Keighley, G., Borsook, H., and Lowy, P. (1959) Effects of erythropoietic stimulating factor on inherited anemia in mice. *Physiologist* **2**: No. 3 (Abstr.).

Russell, E. S., Silvers, W. K., Loosli, R., Wolfe, H. G., and Southard, J. L. (1962) New genetically homogeneous background for dystrophic mice and their normal counterparts. *Science* **135**: 1061-1062.

Russell, J. A., and Bloom, W. (1956) Hormonal control of glycogen in the heart and other tissues in rats. *Endocrinology* **58**: 83-94.

Salcedo, J., Jr., and Stetten, D., Jr. (1943) The turnover of fatty acids in the congenitally obese mouse. *J. Biol. Chem.* **151**: 413-416.

Salt, H. B., Wolff, O. H., Lloyd, J. K., Fosbrooke, A. S., Cameron, A. H., and Hubble, D. V. (1960) On having no beta-lipoprotein. A syndrome comprising a beta-lipoproteinemia, acanthocytosis and steatorrhea. *Lancet* ii: 325-329.

Sarvella, P. A., and Russell, L. B. (1956) Steel, a new dominant gene in the house mouse. *J. Heredity* **47**: 123-128.

Sawin, P. B. (1955) Recent genetics of the domestic rabbit. *Advan. in Genet.* **7**: 183-226.

Sawin, P. B., and Crary, D. D. (1956) Morphogenetic studies of the rabbit. XVI. Quantitative racial differences in ossification pattern of the vertebrae of embryos as an approach to basic principles of mammalian growth. *Am. J. Phys. Anthropol.* **14**: 625-648.

Sawin, P. B., and Crary, D. D. (1957) Morphogenetic studies of the rabbit. XVII. Disproportionate adult size induced by the Da gene. *Genetics* **42**: 72-91.

Sawin, P. B., and Glick, D. (1943) Atropinesterase, a genetically determined enzyme in the rabbit. *Proc. Natl. Acad. Sci. U.S.* **29**: 55-59.

Sawin, P. B., Ranlett, M., and Crary, D. D. (1959a) Morphogenetic studies of the rabbit. XXV. The spheno-occipital synchondrosis of the dachs (chondrodystrophy) rabbit. *Am. J. Anat.* **105**: 257-280.

Sawin, P. B., Crary, D. D., and Webster, J. (1959b) Morphogenetic studies of the rabbit. XXIII. The effects of the dachs gene Da (chondrodystrophy) upon linear and lateral growth of the skeleton as influenced in time. *Genetics* **44**: 609-624.

Schayer, R. W. (1961) Significance of induced synthesis of histamine in physiology and pathology. *Chemotherapia* **3**: 128-136.

Schlichter, J. G., and Harris, R. (1949) The vascularization of the aorta. II. A comparative study of the aortic vascularization of several species in health and disease. *Am. J. Med. Sci.* **218**: 610-615.

Schloesser, C. V. (1886) Acutes Seundaer-Glaucom beini Kaninchen. *Z. vergl. Augent.* **6**: 79-88.

Schlumberger, H. G. (1956) Neoplasia in the parakeet. II. Transplantation of the pituitary tumor. *Cancer Res.* **16**: 149-153.

Schnelle, G. B. (1950) "Radiology in Small Animal Practice," 2nd ed. North Am. Veterinarian, Evanston, Illinois.

Schoenbaum, E., Davidson, M., Large, R. E., and Casselman, W. G. B. (1959) Further studies on the metabolism of glucose and the formation of corticosteroids *in vitro*. *Can. J. Biochem. Physiol.* **37**: 1209-1214.

Schull, W. J., ed. (1962) "Mutations," Second Conference on Genetics. Univ. of Michigan Press, Ann Arbor, Michigan.

Schuman, H. (1955) Die Letalfaktoren bei Hund und Katze. *Berlin. Muench. Tieraerztl. Wochschr.* **68**: 376-378.

Searle, A. G. (1952) A lethal allele of dilute in the house mouse. *Heredity* **6**: 395-401.
Searle, A. G. (1959) Hereditary absence of spleen in the mouse. *Nature* **184**: 1419-1420.
Seegers, W. H., and Alkjaersig, N. (1953) Comparative properties of purified human and bovine prothrombin. *Am. J. Physiol.* **172**: 731-736.
Selye, H. (1958) The humoral production of cardiac infarcts. *Brit. Med. J.* **I**: 599-603.
Selye, H., Veilleux, R., and Cantin, M. (1961) Excessive stimulation of salivary gland growth by isoproterenal. *Science* **133**: 44-45.
Sereni, F., Kenney, F. T., and Kretchmer, N. (1959) Factors influencing the development of tyrosine-α-ketoglutarate transaminase activity in rat liver. *J. Biol. Chem.* **234**: 609-612.
Shapiro, J. R., and Kirschbaum, A. (1951) Intrinsic tissue response to induction of pulmonary tumors. *Cancer Res.* **11**: 644-647.
Sharp, A. A. (1958) Viscous metamorphosis of blood platelets; a study of the relationship to coagulation factors and fibrin formation. *Brit. J. Haematol.* **4**: 28-37.
Sherber, D. A. (1962) Readily extractable cholesterols as an index to species susceptibility to spontaneous atherosclerosis. *Circulation* **24**: 671 (Part 2).
Shock, N. W., ed. (1962) "Biological Aspects of Aging." Columbia Univ. Press, New York.
Shreffler, D. C. (1960) Genetic control of serum transferrin type in mice. *Proc. Natl. Acad. Sci. U.S.* **46**: 1378-1384.
Shubik, P., and Ritchie, A. C. (1953) Sensitivity of male DBA mice to the toxicity of chloroform as a laboratory hazard. *Science* **117**: 285.
Shull, K. H., Ashmore, J., and Mayer, J. (1956) Hexokinase, glucose-6-phosphatase and phosphorylase levels in hereditary obese-hyperglycemic mice. *Arch. Biochem. Biophys.* **62**: 210-216.
Sidman, R. L. (1961) Tissue culture studies of inherited retinal dystrophy. *Diseases Nervous System.* **22** (Suppl.): 1-7.
Sidman, R. L., Perkins, M., and Weiner, N. (1962) Noradrenaline and adrenaline content of adipose tissue. *Nature* **193**: 36-37.
Silberberg, R., and Silberberg, M. (1957) Lesions in "yellow" mice fed stock, high-fat or high-carbohydrate diets. *Yale J. Biol. Med.* **29**: 525-539.
Silverstein, E., and Yamamoto, R. S. (1961) Sex differences in lipid content of adrenal glands in mice. *Proc. Soc. Exptl. Biol. Med.* **106**: 381-383.
Smith, P. E., and MacDowell, E. C. (1930) An hereditary anterior-pituitary deficiency in the mouse. *Anat. Record* **46**: 249-257.
Smith, P. E., and MacDowell, E. C. (1931) The differential effect of hereditary mouse dwarfism on the anterior-pituitary hormones. *Anat. Record* **50**: 85-93.
Snell, G. D. (1929) Dwarf, a new Mendelian recessive character of the house mouse. *Proc. Natl. Acad. Sci. U.S.* **15**: 733-734.
Snell, G. D. (1955) Ducky, a new second chromosome mutation in the mouse. *J. Heredity* **46**: 27-29.
Snyder, J. G., and Wyman, L. C. (1951) Sodium and potassium of blood and urine in adrenalectomized golden hamsters. *Am. J. Physiol.* **167**: 328-332.
Snyder, L. H. (1955) Human heredity and its modern applications. *Am. Scientist* **43**: 391-419.
Sokoloff, L. (1960) Comparative pathology of arthritis. *Advan. Vet. Sci.* **6**: 193-250.
Sokoloff, L., and Barile, M. F. (1962) Obstructive genitourinary disease in male STR/IN mice. *Am. J. Pathol.* **41**: 233-244.

Sokoloff, L., Mickelsen, O., Silverstein, E., Jay, G. E., Jr., and Yamamoto, R. S. (1960) Experimental obesity and osteoarthropathy. *Am. J. Physiol.* **198**: 765-770.

Soloman, J., and Mayer, J. (1962) Effect of alloxan on obese hyperglycemic mice. *Nature* **193**: 135-137.

Sorsby, A., Koller, P. C., Attfield, M., Dave, J. B., and Lucas, D. R. (1954) Retinal dystrophy in the mouse: histological and genetic aspects. *J. Exptl. Zool.* **125**: 171-197.

Soulier, J. P., Wartelle, O., and Ménaché, D. (1959) Hageman trait and PTA deficiency; the rate of contact of blood with glass. *Brit. J. Hematol.* **5**: 121-138.

Speirs, R. S. (1953) Eosinopenic activity of epinephrine in adrenalectomized mice. *Am. J. Physiol.* **172**: 520-526.

Speirs, R. S., and Meyer, R. K. (1949) The effects of stress, adrenal and adrenocorticotropic hormones on the circulating eosinophils of mice. *Endocrinology* **45**: 403-429.

Spoerlein, M. T., and Ellman, A. M. (1961) Facilitation of metrazol-induced seizures by iproniazid and betaphenylisopropylhydrazine in mice. *Arch. Intern. Pharmacodyn.* **133**: 193-199.

Srore, P. A., Chaikoff, I. L., Treitmann, S. S., and Burstein, L. S. (1950) The extrahepatic synthesis of cholesterol. *J. Biol. Chem.* **182**: 629-634.

Staats, J. (1954) A classified bibliography of inbred strains of mice. *Science* **119**: 295-296.

Staats, J. (1958) Behavior studies on inbred mice. A selected bibliography. *Animal Behavior* **6**: 77-84.

Stadie, W. C. (1954) Current concepts of the action of insulin. *Physiol. Rev.* **34**: 52-100.

Steinberg, D. (1962) Chemotherapeutic control of serum lipid levels. *Trans. N. Y. Acad. Sci.* [2] **24**: 704-723.

Stevens, L. C., Mackenson, J., and Bernstein, S. E. (1958) The inheritance and expression of a mutation in the mouse effecting blood formation the axial skeleton, and body size. *J. Heredity* **49**: 153-160.

Stimpfling, J. H. (1960) The variation of heteroagglutinins in normal mouse sera. *J. Immunol.* **85**: 530-532.

Stockard, C. R. (1941) "The Genetics and Endocrine Basis for Differences in Form and Behavior." Wistar Institute, Philadelphia, Pennsylvania.

Stormont, C. (1958) Genetics and disease. *Advan. Vet. Sci.* **4**: 137-162.

Strauss, A., Deitch, A., and Hau, K. (1961) Further observations on the localization of a muscle-binding, complement-fixing serum globulin fraction in myesthenia gravis. *Federation Proc.* **20**: 38.

Takahara, S., Ogata, M., Kobara, T. Y., Nishimura, E. T., and Brown, W. J. (1962) The "catalase-protein" of acetalasemic red blood cells. *Lab. Invest.* **11**: 782-790.

Takayama, S. (1960) Skin carcinogenesis with a single painting of 4-nitroquinoline N-oxide. *Gann* **51**: 139-145.

Tansley, K. (1954) An inherited retinal degeneration in the mouse. *J. Heredity* **45**: 123-127.

Tappel, A. L., Zalkin, H., Caldwell, K. A., Desai, I. D., and Shibko, S. (1962) Increased liposomal enzymes in genetic muscular dystrophy. *Arch. Biochem. Biophys.* **96**: 340-346.

Tavormina, P. A., Gibbs, M., and Huff, J. W. (1956) The utilization of β-hydroxy-β-methyl-δ-valerolactone in cholesterol biosynthesis. *J. Am. Chem. Soc.* **78**: 4498-4499.

Tennent, D. M., Siegel, H., Zanelti, M. E., Kuron, G. W., Ott, W. H., and Wolf, F. J. (1960) Plasma cholesterol lowering action of bile acid binding polymers in experimental animals. *J. Lipid Res.* **1**: 469-473.

Thompson, J. H., Spittel, J. A., Pascuzzi, C. A., and Owens, C. A. (1960) Laboratory and genetic observations in another family with the Hageman trait. *Proc. Mayo Clin.* **35**: 421-427.

Thompson, M., and Mayer, J. (1962) Coenzyme A and acetylation in various experimental obesities. *Am. J. Physiol.* **202**: 1005-1010.

Tindal, J. S. (1960) A breed difference in the lactogenic response of the rabbit to reserpine. *J. Endocrinol.* **20**: 78-81.

Treacher, R. J. (1962) Amino-acid excretion in canine cystine-stone disease. *Vet. Record* **74**: 503-504.

Trimble, H. C., and Keeler, C. E. (1938) The inheritance of "high uric acid excretion" in dogs. *J. Heredity* **29**: 281-289.

Tyler, W. S., Julian, L. M., and Gregory, P. W. (1957) The nature of the process responsible for the short-headed hereford dwarf as revealed by gross examination of the appendicular skeleton. *Am. J. Anat.* **101**: 477-496.

Uzman, L. L., and Rumley, M. K. (1958) Changes in the composition of the developing mouse brain during early myelination. *J. Neurochem.* **3**: 171-185.

van Creveld, S., and Mochtar, I. A. (1959) Management of hemophilia. *In* "Hemophilia and Other Hemorrhagic States" (K. M. Brinkhous and P. De Nicola, eds.), pp. 27-37. The Univ. of North Carolina Press, Chapel Hill, North Carolina.

van Heyninger, H. E. (1961) The initiation of thyroid function in the mouse. *Endocrinology* **69**: 720-727.

Vogel, F. (1959) Moderne Probleme der Humangenetik. *Ergeb. Inn. Med. Kinderheilk.* **12**: 52-62.

Warner, E. D., Brinkhous, K. M., and Smith, H. P. (1939) Plasma prothrombin levels in various vertebrates. *Am. J. Physiol.* **125**: 296-300.

Waterman, A. J., and Gorbman, A. (1956) Development of the thyroid gland of the rabbit. *J. Exptl. Zool.* **132**: 509-538.

Weaver, L. C., and Kerley, T. L. (1962) Strain difference in response of mice to d-amphetamine. *J. Pharmacol. Exptl. Therap.* **135**: 240-244.

Weil-Malherbe, H. (1950) Significance of glutamic acid for the metabolism of nervous tissue. *Physiol. Rev.* **30**: 549-568.

Wells, H. G. (1918) The purine metabolism of the Dalmatian coach hound. *J. Biol. Chem.* **35**: 221.

Welsh, J. F. (1960) Serum transfusion treatment of deficiency of plasma thromboplastic component. *Am. J. Clin. Patholo.* **33**: 118-123.

Werner, S. C., Valpert, E. M., and Grinberg, R. (1961) Difference in metabolism of labelled thyroxine between thyrotropic and adrenotropic mouse pituitary tumors. *Nature* **192**: 1193-1194.

West, W. T., and Murphy, E. D. (1960) Histopathology of hereditary, progressive muscular dystrophy in inbred strains 129 mice. *Anat. Record* **137**: 279-295.

West, W. T., and Murphy, E. D. (1962) Amyloidosis in strain A mice. *Roscoe B. Jackson Memorial Lab. Ann. Rept. No.* **33**: 62-63.

Wexler, B. C., and Miller, B. F. (1958) Severe arteriosclerosis and other diseases in the rate produced by corticotrophin. *Science* **127**: 590-591.

Williams, R. T. (1959) "Detoxification Mechanisms," 2nd ed., p. 721. Chapman & Hall, London.

Williams, R. T. (1962) Enzymes in young animals. *In* "Enzymes and Drug Action,"

Ciba Foundation Symposium (J. L. Mongar and A. V. S. de Reuck, eds.), p. 512. Little, Brown, Boston, Massachusetts.

Wolfe, H. G., and Southard, J. L. (1962) Production of all-dystrophic litters of mice by artificial insemination. *Proc. Soc. Exptl. Biol. Med.* **109**: 630-633.

Wollman, S. H., and Wodinsky, I. (1955) Localization of protein-bound I^{131} in the thyroid gland of the mouse. *Endocrinology* **56**: 9-20.

Woolley, D. W. (1959) Antimetabolites. *Science* **129**: 615-621.

Woolley, D. W. (1962) Alteration in learning ability caused by changes in serotonin or catecholamines. *Science* **136**: 330.

Wooley, G. W. (1948) Further study of the dwarf (dw_2) rat. *Genetics* **33**: 132. (Abstr.)

Yerganian, G. (1958) The striped-back or Chinese hamster (*Cricetulus griseus*). *J. Natl. Cancer Inst.* **20**: 705-735.

Yerganian, G., and Meier, H. (1959) Spontaneous hereditary diabetes mellitus in the Chinese hamster (*Cricetulus griseus*). A preliminary report on clinico-pathological findings, genetic aspects, breeding and responses to hypoglycemic therapy. *Federation Proc.* **18**: 514.

Yoon, C. H. (1961a) Electrophoretic analysis of the serum proteins of neurological mutations in mice. *Science* **134**: 1009-1010.

Yoon, C. H. (1961b) Serum cholinesterase activities of neurological mutants in mice. *Am. Zoologist* **1**: 40 (Abstr.).

Zomzely, C., and Mayer, J. (1959) Endogenous dilution of administered labeled acetate during lipogenesis and cholesterogenesis in two types of obese mice. *Am. J. Physiol.* **196**: 956-960.

Zucker, L. M., and Zucker, T. F. (1961) Fatty, a new mutation in the rat. *J. Heredity* **52**: 275-278.

Zucker, M. B., and Borrelli, J. (1960) Viscous metamorphosis produced by chilling and by clotting failure to find scientific defect of viscous metamorphosis in PTA syndrome. *Thromb. Diath. Haemorrhag.* **4**: 424-434.

Zucker, T. F. (1953) Problems in breeding for quality. *In* "Rat Quality, A Consideration of Heredity, Diet and Disease," pp. 48-70. The National Vitamin Foundation, New York.

Zucker, T. F., and Zucker, L. M. (1962) Hereditary obesity in the rat associated with high serum fat and cholesterol. *Proc. Soc. Exptl. Biol. Med.* **110**: 165-171.

Author Index

The numbers in italics show the page on which the complete reference is listed.

A

Abelman, M. A., 88, *176*
Adams, R. D., 97, *171*
Adamson, R. H., 7, 83, *171, 177*
Adrouny, G. A., 27, *171*
Akamatsu, S., 9, *171*
Alexander, B., 151, *171*
Alkjaersig, N., 156, *191*
Allen, L., 114, *171*
Allen, R. C., 20, 21, 56, 141, 142, 143, 144, 145, 146, 147, 153, 155, 158, 160, *171, 186*
Allison, A. C., 117, *171*
Alpert, M., 96, *185*
Altman, K. I., 67, *171*
Altrocchi, P. H., 168, *178*
Amarger, J., 113, *183*
Ambrus, J. L., 22, *171*
Amin, A., 20, 25, *171, 175*
Ammon, R., 113, *171*
Anderson, W. E., 92, *188*
Andreassen, M., 149, *171*
Anthony, W. L., 91, *172*
Ashmore, J., 39, *191*
Ashton, G. C., 19, *171*
Ashworth, M. E., 66, *175*
Asmundson, V. S., 97, *182*
Assali, N. S., 88, *171*
Attfield, M., 76, *192*
Audibert, A., 113, *183*
Avioli, L. V., 168, *178*
Axelrod, J., 37, 93, *171, 172, 182*

B

Babel, J., 115, *172*
Babikian, L. G., 42, *172*
Bacon, R. L., 96, *183*
Badrick, F. E., 41, *188*
Baker, G. D., 91, *172*
Baker, J. R., 51, 52, 97, *181*
Ball, C. R., 154, *172*

Barile, M. F., 78, *191*
Barnett, R. J., 36, *185*
Barrows, C. H., Jr., 7, *172*
Basser, W., 155, *187*
Bates, M. W., 33, 35, 37, *172, 185*
Bearse, G. E., 129, *184, 186*
Beaton, G. H., 88, *176*
Beck, L. V., 39, *172*
Beher, W. T., 91, *172*
Benditt, E. P., 72, *172*
Benedict, S. R., 119, *172*
Bennet, E. L., 86, 87, *172, 189*
Bennett, D., 68, 73, *172*
Benson, G. K., 113, *172*
Bernick, S., 130, 131, *172, 178, 187*
Bernstein, S. E., 67, *192*
Berthet, L., 39, 41, *180*
Beyer, R. E., 27, *173*
Biancifiori, C., 30, *173*
Bidwell, E., 154, *173*
Bieber, R. E., 8, *183*
Bielschowsky, F., 70, 92, *173*
Bielschowsky, M., 69, 70, 92, *173*
Biggers, J. D., 25, *173*
Biggs, R., 155, 161, *173, 185*
Bigland, C. H., 139, *173*
Bizzi, L., 90, *178*
Black, R., 83, *171*
Blanc, W. A., 93, *182*
Blaschko, H., 84, *173*
Bleisch, V. R., 37, *173*
Bloom, W., 27, *190*
Blount, I. H., 19, *173*
Blount, R. F., 19, *173*
Bogart, P. J., 125
Bogart, R., 149, *173, 181*
Bohren, B. B., *126*
Borges, P., 73, *188*
Borrelli, J., 150, *194*
Borsook, H., 68, *190*
Bosanquet, F. D., 97, *173*

Bowman, B. H., 47, *173*
Boxer, G. E., 65, 66, *173*
Boyland, E., 22, *173*
Braden, A. W. H., 19, *171*
Braley, A. E., 114, *171*
Brinkhous, K. M., 140, 148, 155, *179, 193*
Brobeck, J. R., 92, *173*
Brodie, B. B., 6, 37, 47, 84, *173, 183, 188*
Brown, P. J., 51, *187*
Brown, W. H., 113, *188*
Brown, W. J., 118, *192*
Brown-Grant, K., 133, *173*
Bruell, J. H., 20, *174*
Brummel, E., 41, *188*
Buckwalter, J. A., 148, 155, *179*
Burian, H. M., 114, *171, 174*
Burnet, F. M., 69, 70, *174, 181*
Burnett, W. T., 54, *178*
Burns, J. J., 5, *175*
Burnstein, S., 116, *174*
Burstein, L. S., 88, 117, *187, 192*
Butterworth, K. R., 121, 122, *174*

C

Cahill, G. F., 36, *184*
Caldwell, K. A., 66, *192*
Calkins, E., 121, *174*
Cameron, A. H., 168, *190*
Canal, N., 64, *174*
Cantin, M., 133, *191*
Carbone, J. V., 93, *174*
Caroczy, A. F., 20, *174*
Carroll, F. D., 126, 128, *174, 179*
Carstensen, H., 39, 40, 41, *174, 184*
Carter, T. C., 17
Casals, J., 89, *177*
Caspari, E., 32, *174*
Casselman, W. G. B., 41, *190*
Castle, W. E., 79, 81, 115, *174*
Chai, C. K., 17, 19, 20, 24, 25, 58, 62, 110, *171, 174, 175*
Chaikoff, I. L., 88, 131, *183, 192*
Chambers, D. J., 126, *185*
Chapman, A. B., 85, *183*
Chapman, D., 71, *182*
Chase, H. B., 27, 51, *175, 187*
Chiesara, E., 87, *183*
Christie, N. P., 54, *181*

Christophe, J., 35, 37, 40, *175*
Clare, P., *125*
Clarke, C. A., 8, *181*
Cohen, M. M., 100, *175*
Cole, V. V., 92, *180*
Coleman, D. L., 16, 42, 44, 64, 65, 66, *175, 179*
Colman, R., 151, *171*
Cologne, A., 88, *175*
Connamacher, R. H., 45, *180*
Conney, A. H., 5, *175*
Constantinides, P., 41, *175*
Cook, R. P., 130, *175*
Cooper, J. R., 45, *175*
Cordero Funes, J. R., 20, *184*
Cordova, C. C., 166, *187*
Costa, E., 47, *181, 183*
Courrier, R., 88, *175*
Cowie, A. T., 113, *172*
Craft, W. A., *125*
Craig, C., 92, *186*
Cranston, E. M., 21, *176*
Crary, D. D., 113, 114, *190*
Crenshaw, W. W., 126, *176*
Creveling, C. R., 47, *183*
Crooker, C. M., 51, 52, *181*
Crossland, J., 86, *172*
Csonke, E., 129, *186*
Curry, D. M., 88, *176*

D

Dagenais, Y., 37, 40, *175*
Dagg, C. P., 28, 29, *176, 189*
Daniel, P. M., 97, *173*
Dave, J. B., 76, *192*
Davidson, M., 41, *190*
Davis, B. K., 50, *176*
Davison, A. N., 45, *176*
Day, R., 93, *182*
De Costa, E. J., 88, *176*
Degenhardt, K. H., 110, *175*
Deitch, A., 70, *192*
Denny-Brown, D., 97, *171, 176*
Deol, M. S., 17
Deringer, M. K., 23, 31, *176, 184*
Desai, I. D., 66, *192*
De Schaepdryver, A. F., 123, *176*
Deuel, H. J., Jr., 130, *176*

Author Index

Devlin, T. M., 65, 66, *173*
Dickie, M. M., *18,* 33, 37, *173, 182, 184, 185*
Didisheim, P., 137, 138, 140, 143, *176*
Dieke, S. H., 83, *176*
Di Francesco, A., 151, *171*
Diner, W. C., 121, *174*
Diotti, M. M., 8, *183*
Doell, R. G., 29, *176*
Dole, V. P., 36, *176*
Doolittle, D. P., 14, *179*
Doolittle, R. F., 140, *176*
Dorfman, R. I., 116, *174*
Dowben, R. M., 94, *176*
Dowling, J. E., 76, *176*
Downie, H. G., 149, 151, 155, *187*
Duguid, T. R., 27, *177*
Dunn, T. B., 23, *176*
Dyne, N., 86, 87, *172*
Dyser, T. C., *125*

E

Elliot, F. H., 41, *176*
Ellman, A. M., 46, *192*
Emorson, G., 82, *176*
Engel, F. L., 35, 37, *185*
Eränkö, O., 122, *177*
Ershoff, B. H., 131, *172*
Erspamer, V., 84, *177*
Eschenbrenner, A. B., 31, *184*
Evans, H. M., 96, 128, 131, *177, 179, 183*

F

Faigle, J. W., 136, *177*
Fainstat, T. D., 28, 88, *177, 178*
Falconer, D. S., 58, 68, 93, *177*
Fantl, P., 138, 139, *177*
Fekete, E., 68, *189*
Fellman, J. H., 45, *177*
Fenton, P. F., 27, *177*
Fenwick, M., 41, *188*
Field, R. A., 148, *177, 182*
Figueroa, R. B., 84, *177*
Fisher, M. S., *20*
Fishman, V., 9, *179*
Fishman, W. H., 8, *177*
Folch, J., 89, *177*
Foley, C. W., 126, *177*
Fondal, E. L., 66, 67, *189*

Forsthoefel, P. F., *17*
Fortier, C., 41, *175, 177*
Fosbrooke, A. S., 168, *190*
Fourie, P. J. J., *125*
Fouts, J. R., 5, 7, *177*
Francis, T., 127, *177*
Frankel, H. H., 130, *178*
Frankel, S., 73, *188*
Franks, R. B., 149, *188*
Fraser, F. C., 28, *178*
Frattola, L., 64, *174*
Frederickson, D. S., 168, *178*
Frontino, G., 87, 88, *183*
Fukuoka, F., 30, *187*
Fuller, J. L., 20, 47, *178, 187*
Furth, J., 54, *178*

G

Gadsden, E. L., 54, *178*
Gage, C., 113, *183*
Gaiardoni, P., *178*
Garattini, S., 26, 84, 89, 90, *178*
Gardner, R. E., 82, *185*
Gates, A., 33, *189*
Gerald, P. S., 19, *189*
Geratz, J. D., 155, *178*
Geri, G., 115, *178*
Gibbs, M., 89, *192*
Gilbert, C., 162, *178*
Gilbertson, E., 149, *182*
Gillman, T., 162, *178*
Ginsburg, B. E., 47, *178*
Girard, J., 128, *179*
Glaubach, S., 112, *179*
Glick, D., 112, *179, 190*
Goldberg, M. E., 22, *171*
Goldenberg, H., 9, *179*
Goldstein, S., 22, *171*
Goodman, D. S., 168, *178*
Goodman, H. C., 168, *178*
Goodwin, F., 46, *188*
Gorbman, A., 131, *179, 193*
Gorrie, J., 69, *181*
Goswami, M. N. D., 6, 7, *179*
Gould, A., 64, 65, *179*
Graham, J. B., 148, 155, *178, 179*
Green, D. M., 48, *188*
Green, E. L., 13, 14, *179*

Author Index

Green, M. C., *16, 17, 18, 44, 179*
Green, M. N., 107, 108, *179*
Greenblatt, R. B., 96, *183*
Greenspan, E. M., 19, *184*
Gregory, P. W., *125,* 126, 128, *174, 179, 182, 193*
Grewal, M. S., 69, *179*
Grinberg, R., 55, *193*
Grodsky, G. M., 93, *174*
Grossi, E., 26, 88, 90, *178, 179*
Grossman, M. S., 161, *181*
Grüneberg, H., *17,* 56, 66, 67, 83, 93, *179*
Grunnet, J., 127, *179*
Gunn, C. H., 93, *180*
Gunther, M. S., 27, *175*
Guth, P. S., 22, *171*

H

Hage, T. S., 126, *182*
Halick, P., 156, *180*
Halkerston, I. D. K., 41, *188*
Hamburgh, M., *17,* 58, *180*
Hamilton, C. M., 129, *184*
Hamolsky, M., 8, *183*
Hancock, O., *126*
Hanna, B. L., 114, 115, *180*
Harigaya, S., 9, *171*
Harman, D., 31, *180*
Harman, P. J., 56, 62, *180, 186*
Harned, B. K., 92, *180*
Harris, H., 119, 166, *180*
Harris, R., 130, *190*
Harrison, J. W. E., 22, *171*
Harshfield, G. S., 126, *182*
Harth, S., 56, *185*
Hartley, L. J., 148, 155, *179*
Hathorn, M., 162, *178*
Hattori, K., 137, 138, 140, 143, *176*
Hau, K., 70, *192*
Haynes, R. C., 39, 41, *180*
Hedgardt, G., 155, *187*
Heidenreich, C. J., 126, *177*
Hellerstein, H. K., 20, *174*
Hellman, B., 35, 38, 39, 40, 41, *174, 180, 184*
Helyer, B. J., 69, *173*
Hess, S. M., 45, *180*
Heston, W. E., 12, 23, 30, 31, 82, *176, 180, 181, 184*

Hillarp, N. A., 123, *181*
Hirsch, C. W., 47, *181, 183*
Hoag, W. G., 19, 20, 21, 56, 59, 60, 61, 75, 141, 142, 143, 144, 145, 146, 147, 153, 154, 155, 158, 160, *171, 185, 186*
Hoagland, H., 117, *187*
Hobaek, A., 114, *181*
Hodgkinson, C. P., 88, *185*
Hodgman, B., *126*
Hoeksema, T. D., 149, 151, *187*
Hoffman, R. A., 84, *181*
Hofman, F. G., 54, *181*
Hogan, A. G., 149, *181*
Hollifield, G., 38, *181*
Holmes, M. C., 69, 70, *174, 181*
Homburger, F., 51, 52, 97, 161, *181*
Howie, J. B., 69, *173*
Hrachovy, J. J., 63, *187*
Hrubant, H. E., 73, *181*
Hsia, D. Y. Y., 94, *176*
Hubble, D. V., 168, *190*
Huber, H. L., 129, *181*
Huff, J. W., 89, *192*
Huff, R. L., 42, *181*
Huff, S. D., 24, 42, *181*
Hummel, K. P., 26, 71, 163, *181, 182*
Hunt, H. R., 67, *187*
Hutt, F. B., 128, 148, *177, 182*

I

Ingalls, A. M., 33, *182*
Inscoe, J. K., 93, *182*
Irwin, S., 8, *182*
Isaacson, J. H., 68, 93, *177*
Israels, M. C. G., 149, *182*

J

Jackel, S. S., 83, *187*
Jacquez, J. A., 9, *182*
Jara, Z., 140, *182*
Jay, G. E., Jr., 22, 40, 78, 80, *182, 191*
Jeanrenaud, B., 35, *175*
Jewell, J. S., 88, *185*
Johannesson, T., 57, *125, 182*
Johansson, H., *125*
Johnson, L., 93, *182*
Johnson, L. E., 126, *182*
Jordan, E., 20, 56, 59, 145, *186*

Author Index

Jorgenson, S. K., *125*
Julian, L. M., 97, 113, 126, *179, 182, 193*

K

Kahn, D., 121, *174*
Kalow, W., 8, *182*
Kandutsch, A. A., 52, 53, 54, *182, 183*
Kaplan, N. O., 8, *183*
Karki, N. T., 6, *183*
Karlsson, H., 86, 87, *172*
Kato, R., 87, 88, *183*
Katz, L. N., 130, *183*
Keberle, H., 136, *177*
Kecler, C. E., 56, 119, *183, 193*
Kehl, R., 113, *183*
Keighly, G., 68, *190*
Kelleher, P. C., 75, *183*
Kelly, M. G., 45, *186*
Kelton, E. D., 56, *183*
Kennedy, C., 82, *184, 187*
Kenney, F. T., 6, *191*
Kerley, T. L., 23, *193*
Kimura, Y., 129, *186*
King, C. G., 83, *187*
King, F. J., 47, *173*
King, L. S., 56, *183*
Kirkman, H., 96, *183*
Kirschbaum, A., 29, *191*
Kiyomoto, A., 9, *171*
Kliman, A., 151, *171*
Klotz, A. P., 84, *177*
Knox, W. E., 6, 7, *179, 183*
Kobara, T. Y., 118, *192*
Koller, P. C., 76, *192*
Kominz, D. R., 157, *184*
Koneff, A. A., 96, 131, *183*
Koritz, S. B., 41, *180*
Kreech, D., 86, 87, *172, 189*
Kretchmer, N., 6, *191*
Kruger, F. A., 96, *185*
Kuntzman, R., 6, 47, *181, 183*
Kunze, H., 67, *183*
Kupperman, H. S., 96, *183*
Kuron, G. W., 132, *192*
Kyle, W. K., 85, *183*

L

Lacassagne, A., 20, *184*
Laki, K., 157, *184*

Lambert, W. V., 82, *184*
Lamesta, L., 26, 84, *178*
Landauer, W., 113, *184*
Lane, P. W., 33, 58, *184*
Laney, F., 156, *184*
Large, R. E., 41, *190*
Laroche, M. J., 37, *171*
Larsson, S., 35, 38, 39, 40, 41, *174, 180, 184*
Lasley, J. F., 126, *177*
Lausen, H. H., 57, *182*
Law, L. W., 19, *184*
Lawe, J. E., 106, *184*
Lawson, F. A., 67, 68, *189*
Le Baron, F. N., 89, *177*
Leboeuf, B., 36, 37, *184*
Leboeuf, N., 36, 37, *184*
Lempert, H., 149, *182*
Les, M., 89, *177*
Lessin, A. W., 84, *184*
Levy, J., 112, *184*
Lewis, J. H., 137, 138, 140, 143, *176*
Lillis, D. V., 129, *184*
Liu, S. C. C., 39, *172*
Llach, J. L., 20, *184*
Lloyd, J. K., 168, *190*
Lochaya, S., 36, 37, *184*
Long, C. N. H., 92, *173*
Loosli, R., 62, *184, 190*
Lorand, L., 157, *184*
Lorenz, E., 31, *184*
Lowy, P., 68, *190*
Lucas, D. R., 76, *184, 192*
Luce, S. A., 89, *184*
Lucey, J. F., 93, *182*
Luecke, R. W., 82, *184*
Lyon, M. J., 16

M

Maas, J. W., 32, *185*
McClearn, G. E., 19, 83, *185*
McCone, W., 126, *182*
MacDowell, E. C., 127, *191*
Macfarlane, R. G., 155, 161, *173, 185*
McGaughey, C., 74, *185*
Mackenson, J., 67, *192*
McLaren, A., 25, *173*
Madjerek, Z., 133, *185*

Maickel, R. P., 37, *188*
Malaragno, H. P., 51, *187*
Mandel, P., 56, *185*
Mann, M., 121, 122, *174*
Margulis, R. R., 88, *185*
Marks, B. H., 96, *185*
Marlowe, T. J., 126, *185*
Marquardt, W., 114, *185*
Marshall, N. B., 35, 36, 37, *185*
Martin, G. J., 82, *185*
Martinez, A., 56, *185*
Mauss, S. F., 35, *172*
Mayer, J., *17*, 33, 35, 36, 37, 38, 39, 40, 92, *172, 173, 175, 184, 185, 191, 192, 193, 194*
Mead, S. W., *125*
Meath, J. A., 89, *177*
Meier, H., 11, 19, 20, 21, 22, 27, 33, *34, 36, 37*, 46, 47, 56, 59, 60, 61, 75, 97, 98, 100, 104, 106, 107, 108, 141, 142, 143, 144, 145, 146, 147, 153, 154, 155, 158, 160, *171, 179, 185, 186*
Meites, J., 113, *186*
Melcer, I., 45, *175*
Ménaché, D., 138, *192*
Mengel, C. E., 45, *186*
Merskey, C., 149, *186*
Meyer, R. K., 26, *192*
Michael, E., 112, *184*
Mickelsen, O., 40, 78, 92, *186, 191*
Michelson, A. M., 16, 62, *186*
Miller, B. F., 162, *193*
Miller, J., 27, *175*
Miller, K. D., 156, *186*
Miller, V. L., 129, *184, 186*
Mintz, B., *18*, 68, *186*
Mirand, E. A., 127, *186*
Mitchie, D., 25, *173*
Mixter, R., 67, *187*
Mochtar, I. A., 154, *193*
Mollenback, C. J., 127, *187*
Montagna, W., 51, *187*
Mordkoff, A. M., 20, *187*
Mori, K., 30, *187*
Morris, H. P., 82, *187*
Morrow, A. G., 19, *184*
Mosbach, E. H., 83, *187*
Mostari, A., 26, 84, *178*

Motulsky, A. G., 1, 8, *187*
Muhrer, M. E., 149, *173, 181*
Murphy, E. D., 62, 64, 72, *193*
Mustard, J. F., 149, 151, 155, *187*

N

Nachmias, V. T., 44, *187*
Nachtsheim, H., 2, *187*
Nadel, E. M., 116, 117, *174, 187*
Naidoo, S. S., 162, *178*
Nakahara, W., 30, *187*
Nathaniel, D. R., 130, *187*
Nichols, C. W., Jr., 131, *183*
Nishimura, E. T., 118, *192*
Niu, L. C., 166, *187*
Niu, M. C., 166, *187*
Nixon, C. W., 97, *181*
Noell, W. K., *16, 17*
Nour-Eldin, F., 155, *187*

O

Ogata, M., 118, *192*
Ohlander, A., 86, 87, *172*
Ohshima, S., 9, *171*
Osborne, C. M., 127, *186*
O'Steen, W. K., 63, *187*
Ott, W. H., 132, *192*
Owens, C. A., 138, *193*

P

Paigen, K., *16, 17*, 19, *188*
Palma, V., *178*
Palmer, L. S., 82, *184, 187*
Paoletti, P., 88, 89, *178, 179, 188*
Paoletti, R., 26, 37, 88, 89, 90, *178, 179, 187*
Pare, C. M. B., 45, *188*
Parkes, M. W., 84, *184*
Parry, H. B., 97, *173*
Parson, W., 38, *181*
Pascuzzi, C. A., 138, *193*
Patek, P. R., 130, 131, *172, 178, 187*
Pattengale, P. S., 126, *179*
Pearce, L., 113, *188*
Pearson, C. M., 97, *171*
Perkins, A., 47, *178*
Perkins, M., 37, *191*
Perlman, M., 38, *181*

Author Index

Peron, F. G., 41, *180*
Pickens, M., 92, *188*
Pilling, J., 161, *173*
Pinninger, J. L., 149, *188*
Pipes, G. W., 126, *176*
Pitt-Rivers, R., 131, *188*
Plotnikoff, N. P., 48, *188*
Poel, W. E., 101, *188*
Poggi, M., 26, 90, *178*
Poiley, S. M., 79, *188*
Ponsetti, I. V., 114, *174*
Pope, A., 89, *177*
Popp, D. M., 57, *188*
Popp, R. A., 57, *188*
Preziosi, P., 123, *176*
Price Evans, D. A., 8, *188*
Putilin, K. I., 118, *188*

Q

Quick, A. J., 149, *188*

R

Radbill, C. L., 166, *187*
Radev, T., 117, *188*
Rankin, R. M., 131, *188*
Ranlett, M., 114, *190*
Rattan, C. P., 88, *185*
Reed, J., 83, *188*
Reineke, E. P., 20, 25, *171, 175*
Reiss, M., 41, *188*
Renold, A. E., 35, *175*
Renson, J., 46, *188*
Ribacchi, R., 30, *173*
Richter, C. P., 83, *176*
Rickard, C. G., 148, *177, 182*
Riess, W., 136, *177*
Ritchie, A. C., 23, *191*
Roberts, E., 47, 62, 73, 82, *175, 178, 188*
Robinson, G. A., 149, *187*
Robinson, R., 95, *189*
Roderick, T. H., 19, 85, *189*
Roeder, L. M., 7, *172*
Rogers, D. A., 19, 83, *185*
Rogers, S., 29, *189*
Rollins, W. C., 126, *174, 179*
Rosen, F., 22, *189*
Rosengren, E., 21, *189*
Rosenzweig, M. R., 86, 87, *172, 189*

Ross, S., 47, *178*
Roswell, H. C., 149, 151, 155, *187*
Roth, N., 128, *179*
Rumley, M. K., 89, *193*
Runner, M. N., 29, 33, *189*
Ruppert, H. L., Jr., 126, *176*
Russell, A. E., 52, *182, 183*
Russell, E. S., 14, 15, *18,* 19, 62, 63, 66, 67, 68, *171, 184, 186, 189, 190*
Russell, J. A., 27, *171, 189*
Russell, L. B., *17,* 67, *190*

S

Salcedo, J., Jr., 92, *190*
Salomon, K., 67, *171*
Salt, H. B., 168, *190*
Sandler, M., 45, *176, 188*
Sarmiento, F., 93, *182*
Sarvella, P. A., *17,* 67, *190*
Sawin, P. B., 109, 111, 112, 113, 114, 115, *180, 190*
Schabtach, G., 67, *189*
Schally, A. V., 41, *176*
Schayer, R. W., 21, *190*
Schlichter, J. G., 130, *190*
Schloesser, C. V., 114, *190*
Schlumberger, H. G., 101, *190*
Schmid, K., 136, *177*
Schnelle, G. B., 120, *125, 190*
Schoenbaum, E., 41, *190*
Scholtz, E., 151, *171*
Schull, W. J., 134, *190*
Sciuchetti, A. M., 82, *184*
Scott, H. M., 126
Scott, J. K., 67, *171*
Searle, A. G., *16,* 58, *191*
Secord, D., 151, 155, *187*
Seegers, W. H., 156, 157, *180, 184, 186, 191*
Selye, H., 133, 162, *191*
Sereni, F., 6, *191*
Shapiro, J. R., 29, *191*
Sharp, A. A., 150, *191*
Sheppard, L. B., 114, 115, *180*
Sherber, D. A., 130, *191*
Sherman, J. H., 9, *182*
Shibko, S., 66, *192*
Shklar, G., 100, *175*

Shock, N. W., 7, *191*
Shore, P. A., 84, *173*
Shreffler, D. C., 19, *191*
Shubik, P., 23, *191*
Shull, K. H., 39, *191*
Sidman, R. L., 37, 62, 76, *175, 19_*
Siegel, H., 132, *192*
Silberberg, M., 33, *191*
Silberberg, R., 33, *191*
Silvers, W. K., 62, *184, 190*
Silverstein, E., 20, 40, 78, *191*
Simpson, M. E., 128, *177, 183*
Sirlin, J. L., 56, *185*
Skeleton, F. R., 41, *175*
Smith, A. H., 92, *188*
Smith, H. P., 140, *193*
Smith, L. J., 68, *189*
Smith, P. E., 127, *191*
Smith, R. L., 37, *188*
Snell, G. D., *16,* 33, 58, 127, *182, 191*
Snyder, J. G., 96, *191*
Snyder, L. H., 1, *191*
Sokoloff, L., 40, 78, *191*
Soloman, J., 38, *192*
Sorsby, A., 76, *192*
Soulier, J. P., 138, *192*
Southard, J. L., 62, 66, *184, 190, 194*
Spector, S., 84, *173*
Speirs, R. S., 26, *192*
Spittel, J. A., 138, *193*
Spoerlein, M. T., 46, *192*
Srore, P. A., 88, *192*
Staats, J., 19, *192*
Stacey, R. S., 45, *188*
Stadie, W. C., 40, *192*
Stamler, J., 130, *183*
Steinberg, D., 132, *192*
Steinhardt, C., 21, *189*
Stetten, D., Jr., 92, *190*
Stevens, L. C., 67, *192*
Stimpfling, J. H., 19, *192*
Stockard, C. R., 113, *192*
Storer, J. B., 19, *189*
Strauss, A., 70, *192*
Sugimura, T., 30, *187*
Surgenor, D. M., 140, *176*
Suyemoto, R., 88, *171*
Symonds, P., 157, *184*

T

Takahara, S., 118, *192*
Takahashi, S., 92, *186*
Takayama, S., 30, *192*
Tansley, K., 56, *192*
Tappel, A. L., 66, *192*
Tavormina, P. A., 89, *192*
Tennent, D. M., 132, *192*
Tepperman, J., 92, *173*
Thomson, R. D., 130, *175*
Thompson, J. H., 138, *193*
Thompson, M., 35, *193*
Tindal, J. S., 113, *172, 193*
Tomchick, R., 37, *172*
Tramezzani, J. H., 20, *184*
Treacher, R. J., 120, *193*
Tregier, A., 51, 52, 161, *181*
Treitmann, S. S., 88, *192*
Triantaphyllopoulos, D. C., 139, *173*
Trimble, H. C., 119, *193*
Turner, C. W., 126, *176*
Tyler, W. S., 126, *179, 182, 193*

U

Udenfriend, S., 45, 46, *180, 188*
Uzman, L. L., 89, *193*

V

Valpert, E. M., 55, *193*
Valzelli, L., 84, *178*
van Creveld, S., 154, *193*
van der Vries, J., 133, *185*
van Heyninger, H. E., 131, *193*
Vassanelli, P., 87, 88, *183*
Veilleux, R., 133, *190*
Vertua, R., 26, 90, *178*
Vest, M., 128, *179*
Villee, C. A., 75, *183*
Vlahakis, G., 30, *181*
Vogel, F., 1, *193*
von Noorden, G. K., 114, *174*

W

Wald, G., 76, *176*
Warner, E. D., 140, *193*
Wartelle, O., 138, *192*
Waterman, A. J., 131, *193*

Author Index

Waugh, D. F., 156, *184*
Weaver, L. C., 23, *193*
Webster, J., 114, *190*
Weil-Malherbe, H., 45, 47, *193*
Weiner, N., 37, *191*
Weissbach, H., 46, *188*
Welsh, J. F., 155, *193*
Wells, A., 131, *172*
Wells, H. G., 119, *193*
Werner, S. C., 55, *193*
Werz, G., 113, *171*
West, W. T., 62, 64, 72, *193*
Westman, S., 35, 38, *180*
Wexler, B. C., 162, *193*
Whitney, R., 97, *181*
Wilkinson, J. F., 155, *187*
Williams, R. T., 6, 83, *193, 194*
With, T. K., *125*
Wodinsky, I., 131, *194*
Wolf, F. J., 132, *192*
Wolfe, H. G., 62, 66, *190, 194*
Wolff, J., 131, *183*
Wolff, O. H., 168, *190*
Wolffson, D., 27, *175*
Wollman, S. H., 131, *194*
Wood, F. C., Jr., 36, *184*
Wooley, G. W., 82, *194*
Woolley, D. W., 45, 47, *194*
Wyman, L. C., 96, *191*

Y

Yamamoto, R. S., 20, 40, 78, *191*
Yasuno, A., 30, *187*
Yerganian, G., 33, 95, 97, 98, 100, 101, 104, 106, 107, 108, *175, 179, 186, 188, 194*
Yoon, C. H., 57, 59, *194*
Young, A., *126*

Z

Zalkin, H., 66, *192*
Zamis, J. J., 47, *178*
Zomzely, C., 35, *194*
Zucker, L. M., 91, 92, *194*
Zucker, M. B., 150, *194*
Zucker, T. F., 82, 91, 92, *194*

Subject Index

A

AA('s) *see* Amino acid(s)
2-Acetamidofluorene, metabolism in rats, 9
Acetanilide,
  increased hydroxylation in benzpyrene-treated rats with hereditary jaundice, 93
  metabolism in newborn rabbits, 7
Acetazolamide, carbonic anhydrase inhibition by, 115
Acetoacetate, accumulation in muscular dystrophy of mice, 65
Acetylcholine, concentration in brain of rat strains, 85
Acetylcoenzyme A, hepatic and renal elevation in obese-hyperglycemic mice, 35
N-Acetylsulfonilamides, metabolism in chickens, 9
Adenosine monophosphatase, increased activity in muscular dystrophy of mice, 64
Adenosine triphosphatase, Na-K, inhibition by digitalis alkaloids, 115
  activity in recessive buphthalmos of rabbits, 115
Adenosine triphosphate, relation to stored catecholamines, 123
ACTH *see* Adrenocorticotropic hormone
Adrenal cortical tissue, accessory, in mice (C57, C58, BALB/c, RIII, CE, A, C3H), 26
  influence on responses to ACTH, histamine, cortisone, serotonin, and reserpine, 26, 27
  effect on drug toxicity, 84
Adrenalectomy, enhancement of serotonin and reserpine toxicity by, in rats, 84, 85
  strain differences, increase in toxicity of SU3118 and SU5171
  in rats following, 85
Adrenal cortical tumors in (DBA/2, WyDi × CE/WyDi) F_1 mice, reduced enzymic efficiency in steroid production and conversions, 54

Adrenocorticotropic hormone, stimulation of (adrenal) phosphorylase by, 39
Adrenotropic pituitary tumors, control of secretions in, of mice, 54
Age, influence on drug metabolism, 6, 7
Aging and spontaneous cancer in mice, 31, 32
  action of antioxidants on, 31, 32
AHF (Antihemophilic Factor) *see* Hemophilia A
Alcohol, preference in mice (C57BL/6, DBA/2), 19
  in rats (Wistar, G-4), 83
Alcohol dehydrogenase, 19, 84
Aldosterone, conversion of progesterone-4-C^{14} to, 41
Alkylating agents, potential mutagenicity in man and animals, 134-136
Alloxan diabetes, decrease of hexokinase and glucose-6-phosphatase, 39
  difference in pathogenesis of hyperglycemia, compared with obese-hyperglycemic syndrome in mice, 39
Amino acid(s), metabolism relative to immunologic maturity, 6
  synthesis from glucose in obese-hyperglycemic mice, 41, 42
Aminoacidemia, genotypic difference in mice, 72, 73
Aminoaciduria, in dogs with cystine-stone disease, 119, 120
  genotypic differences in, of mice, 72, 73, 74, 78
  urinary calculi formation in dystrophic mice due to, 74
γ-Aminobutyric acid, 45
2-Aminofluorence, influence on reticular tissue in mice with autoimmune hemolytic anemia, 70
2-Amino-5-nitrothiazole, as inhibitor of estrus in mice, 21
Aminopyrine, metabolism in newborn rabbits, 7

Subject Index

Aminoxyacetic acid, effectiveness in spastic mice, 62
AMPase *see* Adenosine monophosphatase
Amphenone-B [3,3-bis(*p*-aminophenyl)-butanone-2-dihydrochloride], interference with steroidogenesis, 96
  species difference in action of, 95, 96
Amphetamine, antagonists of, 23
  metabolism in newborn rabbits, 7
  response to, in mice (C57BL/6, DBA/2, BDF_1), 23
Amprolium (1[(4-amino-2-*n*-propyl-5-pyrimidinyl)methyl]-2-picolinium chloride), as chicken coccidiostat, 129
Amyloidosis, in A/J mice, 71, 72
  pathogenesis of, role of tanning systems in, 72
Anemia, hereditary, in mice, 66, 67, 68, 69
  analyses of responsible genes, 66, 67, 68, 69
  autoimmune hemolytic, in NZB mice, 69-70
Animals, influence of genetic factor on drug metabolism, 5, 6, 7, 8, 9
Antipyrine, metabolism in rats, 83
Antioxidants, influence on aging process and spontaneous cancer incidence in mice, 31, 32
Ascorbic acid, influence on life span and spontaneous cancer incidence in mice, 31
  inability of guinea pigs to synthesize, 116
  stimulation of excretion by chloretone in rats, 83
Atherosclerosis, relation to cholesterol metabolism, 130, 131, 132
  REC as index of, 130
  to therapeutic approaches, 132
  to thyroid function, 131
ATP *see* Adenosine triphosphate
ATPase *see* Adenosine triphosphatase
Atropine, enzyme splitting, in rabbits, 112
Atropinesterase, genetic and pharmacologic properties in rabbits, 112
Audiogenic seizures (epilepsy) in rabbits, 115, 116
  in dilute mice, 42, 43, 44, 45, 46, 57, 48
  and drug resistance, 47

anticonvulsant action of $d,1$-α-ethyl-tryptamine (Monase), 46, 47
  inhibition of decarboxylases by phenylacetic and pyruvic acids, 45
Auricular thrombosis left, in mice, 152, 153, 154
Autoimmune hemolytic anemia *see* Anemia
Azaserine, detoxification in mice, 9

B

Benzmalacene, hypocholesterolemic effect in suckling rats, 90
Benzpyrene, effect on acetanilide hydroxylation in rats with hereditary jaundice, 93
Binder-protein of insulin, 167
Biphenylbutyric acid, effect on hypercholesterolemia of suckling rats, 90
Blood coagulation, comparative aspects of, 137-163
  in lower vertebrates and birds, 137-140
  in mammals, 140-142
  inhibitors of, 143
Borneol, influence on β-glucuronidase, 9
Breeds and varieties of chickens, 128, 129
Buphthalmos, recessive, in rabbits, 114, 115

C

Caffeine, potential mutagenicity in man and animals, 135
Carbonic anhydrase, activity in estrus cycle, 132
  in assays of progestational activity, species differences in, 132, 133
  inhibition by acetazolamide, 115
Carisopodol, sex-dependent action in rats (Sprague-Dawley), 87
Catalase, genetic control in cattle, dogs, guinea pigs, and man, 117, 118
Catecholamines,
  contents of, in adrenals of families of cats, 121, 122
  in brain of newborn guinea pigs, 6
  drug action on tissue content in mice, 123, 124
  relationship to ATP, 122, 123
  release of, by syronsingopine, 39

species-dependent secretion of, 121-124
  storage of, in animals, 121-124
Catechol-O-methyltransferase, inhibition by pyrogallol and guercetin, 37
Catheptic activity, increase in senescence (rats), 7
Chinese hamsters, spontaneous hereditary diabetes mellitus in, 97-109
Ceruloplasmin (oxidase), effect of adrenalectomy on, in rats, 84
  as "tanning" system in amyloidosis, 72
Chickens, breeds and varieties of, 128, 129
Chloranystenicol, excretion in rats, 9
Chloretone (trichlorobutanol) effects in rats (Wistar, Sherman), 83
  stimulation of ascorbic acid and glucuronide excretion in rats by, 83
Chloroform, toxicity in mice (C3H, C3Hf, A, HR, DBA, C57BL, C57BR/cd, C57L and ST), 23
Chlorinsondiamine diethylchloride, hypoglycemic action in mice, 39
Chlorpromazine (2-chloro-10-[3-(dimethylamino)propyl]phenothiazine)
  action on catecholamine storage in mice, 124
  inhibition of estrus in mice, 21
  metabolism, in dogs and man, 9
  sensitivity to, in mice (C57BL/6J, DBA/2J, A/HeJ, C3HeB/FeJ), 23, 24
Cholesterol, in dilute mice, 89
  relation to atherosclerosis, 130
  synthesis of, in growing rat brain (myelation), 88, 89, 90
  in preputial gland tumor, ESR586, of C57BL/6J mice, 53
Cholesterolemia see hypercholesterolemia
Cholinesterase, activity of, and strain differences in rats, 85, 86, 87
  cryptic form of, in reeler mutant of mice, 58
  suxamethonium sensitivity due to lack of, 167
Chondrodystrophy, hereditary, in rabbits, 113, 114
Copper, retention in chickens, 129
Chorionic gonadotropin, response of mice (DBA/1J, DBA/2J, A/J, and BALB/cJ) to, 25

Coagulation proteins, distribution in mouse plasma, 157, 158, 159
  separation and purification of, 159, 160
Cocaine-esterase, in rabbits, breed difference of, 112, 113
Coccidiostats in chickens, 128, 129
Compound A see 11-Dehydroxycorticosterone
Coronary disease, sexual dimorphism in rats, 161, 162
Corticosteroids, urinary excretion in normal and scorbutic guinea pigs, 116, 117
Corticosterone, principal steroid produced by adrenals of obese-hyperglycemic mice, 40
  conversion from progesterone-4-C^{14}, 40
Cortisone, influence by accessory adrenal cortical tissue on response of, in mice, 26
  resistance of pregnant rats to, 88
Cysteine hydrochloride, influence on life span and spontaneous cancer incidence in mice, 31
Cysteine-stone disease, in dogs, 119, 120
Cystinuria, in man, genetic analysis of, 119, 120
Cystolithiasis, in STR/IN mice, 78

D

Dalmatian dogs, purine metabolism and mechanism of high uric acid excretion in, 119
Decarboxylases, pyridoxine-dependent, inhibition by phenylacetic and pyruvic acids, 45
11-Dehydroxycorticosterone, influence on lipogenesis in obese-hyperglycemic mice, 38
Demethoxyreserpine, action on catecholamine storage in mice, 124
Desmosterol, elevation in suckling rats, 90
Diabetes insipidus, in MA/My and MA/J mice, 70, 71
Diabetes mellitus, hereditary, in Chinese hamsters, 97-109
2,2′-Diaminodiethyl disulfide, influence on lifespan and spontaneous cancer in mice, 31, 32
Dibenz[a,h]anthracene, tumor induction in mice, 31

Subject Index

Difenesemic acid, hypocholesterolemic effects in suckling rats, 90
Differences between inbred strains of mice, 19, 21
Digitalis alkaloids, inhibition of Na-K ATPase by, 115
Dilantin, ineffectiveness in spastic mice, 62
Dilute mice, inhibition of phenylalanine hydroxylase in, 44
  influence of dilution on isoniazid-induced pulmonary tumors, 29, 30
Dimorphism, sexual, in mice, 20, 21
  in adrenal lipid content, 20
Dominant hemimelia, hereditary absence of spleen in mice, 74, 75
Drug metabolism, factors influencing, 5-9
  inhibitors in baby rabbits, 7
Dwarf mutants, similarity of rat and mouse dwarf mutants, 127, 128
  non-homology of cattle and mouse, 126
Dystrophy, *muscular,* in mice, 62-66
  acetoacetate accumulation, 64, 65
  AMPase, increase in activity of, 64
  comparative pathological studies, 62
  inability to utilize 11-*cis*-retinene, 76
  intracellular hydrogen transport, 65
  primary gene effects and metabolic derangements, 64
  pyruvate, defective utilization of, 64, 65
  renal glycine transamidinase elevation, 64
  retinal, inheritance of, in mice, 75-77
  urinary calculi formation, 74

E

Endocrine variation and function, 20, 21, 24-28
Enzymes, alcohol dehydrogenase, 19, 84
  Na-K ATPase, 115
  atropinesterase, 112
  carbonic anhydrase, 115, 132, 133
  catalase, 117, 118
  catechol-O-methyltransferase, 37, 137
  catheptic, 7
  cholinesterase, 58, 85-87, 167
  cocaine-esterase, 112, 113
  dihydrophenylalanine decarboxylase, 45
  esterase, 57
  glucose-6-phosphatase, 39, 166
  β-glucuronidase, 9
  glucuronyl transferase, 93
  glutamic acid dehydrogenase, 45
  glycine transamidinase, 64
  hexokinase, 39
  histidine decarboxylase, 20
  17-α-hydroxylase, 41
  C-11 and C-12 hydroxylase, 54
  p-hydroxyphenylpyruvate oxidase, 6
  5-hydroxytryptophan decarboxylase, 46
  "insulinase," 27
  lactic dehydrogenase, 87
  monoamine oxidase, 84, 85
  phenylalanine hydroxylase, 44, 166
  phosphatase, alkaline, 57
  phosphorylase, 41
  Δ^6-reductase, 53
  succinoxidase, 6, 7
  tyrosinase, 44, 166
  tyrosine transaminase, 6
  tryptophan pyrrolase, 166
Enzyme concentration, senescent changes in rats, 7
  catheptic activity, increase in rats, 7
  succinoxidase, decrease in liver and adrenal cortex of rats, 7
Enzyme inhibitors, in drug metabolism of baby rabbits, 7
  acetazolamide (carbonic anhydrase), 115
  digitalis alkaloids (Na-K ATPase), 115
  glucosaccharo-14-lactose (β-glucuronidase), 9
  guercetin and pyrogallol (catechol-O-methyltransferase), 37, 137
  NSD1034[N-(m-hydroxybenzyl)-N-methylhydrazine] and α-methyl-m-tyrosine, 47
  phenylacetic and phenylpyruvic acid (decarboxylases), 45, 46
  m-tyrosine, 47
Epilepsy, in rabbits, *see* Audiogenic seizures
Epinephrine, relation to adrenal cortical function, 26
Estrogen, induction of malignant renal tumors in male golden hamsters, 96, 97
  influence on hairless mice, 51, 52
Estrus, anti-estrus action in mice of chlorpromazine, meprobamate, nidroxyzone,

perphenazine, 2-amino-5-nitrothiazole, 21
Ethisterone, influence on hairless mice, 51
Ethionine, ineffectiveness in hypercholesterolemia of suckling rats, 90
Excretion, of amino acids, 72-74, 78
of chloranystenicol, in rats, 9
of corticosteroids in normal and scorbutic guinea pigs, 116, 117
of 5-hydroxyindoleacetic acid, in phenylketonuric mice, 44, 45
of morphine, in dogs, 9
of uric acid in Dalmatian dogs, 119

F

Factor, specific, genetically determined deficiencies of blood coagulation, 143-152
AHF, 148, 149
Hageman factor, cat, 143, 144
PTC, 149, 150
Stuart-Prower factor, mice, 144-148
VII, 151, 152
Farm animals, inherited traits of, 124-129
Fibrinolysis, factors influencing, 161-163
fibrinolytic activity of endometrial secretions, 160, 161
FFA see Free fatty acid(s)
5-Fluorouracil, teratogenic action in mice (129, BALB/c), 28
Free fatty acid(s), metabolism of glucose and of, in obese-hyperglycemic mice, 36
Formaldehyde, potential mutagenicity in man and animals, 134

G

GABA see γ-Aminobutyric acid
Gene effects, 15
Genetic factors, dependence of drug metabolism upon, 8, 9
Genitourinary disease, in STR/IN mice, 78
Glucagon, 37
Glucosaccharo-14-lactose, inhibition of β-glucuronidase by, 9
Glucose, incorporation into glycogen and lipids in obese-hyperglycemic mice, 34
Glucose-6-phosphatase, decrease in alloxan diabetes, 38
Primaquine sensitivity due to lack of, 167
RNA-induced synthesis of, 166

β-Glucuronidase, estrogen-dependence of, 9
genetic control in mice (C3H, A), 19
influence of menthol and borneol on, 9
Glucuronide, excretion, stimulation by chloretone in rats, 83
Glucuronyl transferase, deficiency in hereditary non-hemolytic jaundice of rats (Gunn), 93
Glutathione, hyperglycemic action in obese-hyperglycemic mice, 39
Glutamic acid decarboxylase, in phenylketonuric mice, 45
Glycarbylamide, as coccidiostat in chickens, 129
Glycogen, cardiac, relative to growth hormone activity in mice (A,I), 27
hepatic, elevation in obese-hyperglycemic mice, 36
Glycine transamidinase, renal, elevation in muscular dystrophy of mice, 64
Golden hamsters, hereditary polymyopathy in, 97
influence of amphenone in, 95, 96
malignant renal tumors in male, 96, 97
Goldthioglucose, induction of obesity in mice by, and comparison to hereditary obesity in mice, 36
Gonadectomy, influence in mice, 21
Growth hormone, relation to cardiac glycogen in mice, 27
GSH see Glutathione
Guercetin, inhibition of catechol-O-methyltransferase by, 137
Guinea pigs, corticosteroid excretion in normal and scorbutic, 116, 117
hereditary characteristics of, 116
inability to synthesize ascorbic acid, 116
L-Gulonolactone, lack of conversion to ascorbic acid in guinea pigs, 116

H

Hageman factor, deficiency of, 138, 143, 144
Hair growth, endocrine states and drug metabolism relative to, 50
Hairless mice, hormonal influence on skin of, 50
vitamin A metabolism in, 49, 50
rats, genetic defect of, 82

Subject Index

Heliotrin (pyrrolizidine alkaloid), as a potential mutagen, 134
Hemoglobin, physico-chemical properties in mice, 19
Hemolytic anemia, autoimmune, in NZB mice, 69, 70
Hemimelia, dominant, protein production in mice with, 74, 75
Hemophilia A, 148, 149
Hemophilia B, 149-152
Heparinoid, lack of hypocholesterolemic effect in suckling rat, 90
Hepatic microsomes, drug metabolism of, 5
Heredity, mechanism of, 15
Hexestrol, lack of hypocholesterolemic effect in suckling rats, 90
Hexobarbital, metabolism in newborn rabbits, 7
    toxicity in mice (SWR/HeN, A/LN), 22
Hexokinase, decrease in alloxan diabetes, 39
Histamine, toxicity in mice (ICR, C3H/J), 22
    relative to accessory adrenal cortical tissue, 26
Histidine decarboxylase, sexual dimorphism in mice, 20
Homatropine, non-specific rabbit atropinesterase substrate, 112
Hormones, response of inbred and hybrid mice to, 24-27
5HT see Serotonin
5HTP see 5-Hydroxytryptophan
Hybrids, F_1 definition of, 14
Hydrogen peroxide, as a potential mutagen, 134
5-Hydroxyindoleacetic acid, excretion in mice, 45
17-α-Hydroxylase, deficiency in mouse adrenals
p-Hydroxyphenylpyruvate oxidase, in the newborn rat, 6
Hydroxylamine, ineffectiveness in spastic mice, 62
    influence on life span and spontaneous cancer in mice, 31
5-Hydroxytryptophan decarboxylase, in phenylketonuric mice, 46

C-11, C-12 Hydroxylase, reduced activity in adrenal cortical tumors of mice, 54
Hypercholesterolemia, in a rat "fatty" mutant, 91-93
    in obese-hyperglycemic mice, 20, 35
    screening of hypocholesterolemic drugs in suckling rats, 90, 91
    in suckling rats, 90, 91
5 Hydroxytryptophan, competitive decarboxylation of, and DOPA, by α-methyl-m-tyrosine, 47
Hypercorticism, in obese-hyperglycemic mice, 39, 40
Hyperglycemia, in mice with obese-hyperglycemic syndrome, 38, 40
    due to hypercorticism, 39, 40
    in diabetes mellitus of Chinese hamsters, 97-109
    in parakeets, 101
Hyperlipemia, in a rat "fatty" mutant, 91-93
Hypocholesterolemia, effects of drugs, 90, 91
Hypoinsulinism, in diabetes mellitus of Chinese hamsters, 98

I

Inbred strains, definition of, 14
    inbred-pedigreed, 14
    inbred-derived, 14
Inhibitors of enzymes, 7, 9, 37, 45, 46, 115, 137
Insulin, NPH, in normal and diabetic Chinese hamsters, 102-104
    in obese-hyperglycemic mice, 38-40
    resistance to, in KL mice, 27, 28
"Insulinase," cause of insulin resistance in KL mice, 27
Intracellular hydrogen transport, 65
Iproniazid, toxicity in mice (AKR, C57BL, DBA/2, dilute), 22
Isoniazid, metabolites of, causing pulmonary tumors in mice (BALB/c and dilute), 29, 30
Isopropylnoradrenaline, enlargements of salivary glands due to, in mice and rats, 133
Isoproterenol see Isopropylnoradrenaline

J

Jaundice, hereditary non-hemolytic, of rats (Gunn), 93, 94

K

Keto-steroids, excretion in mice with preputial gland tumor, 53
Kidney, elevated glycine transamidinase in, of dystrophic mice, 64
Kinurenine and Kinurenic acid, excretion in mice, 45

L

Lactic dehydrogenase, activity in brain of rats, 87
Lactogenic response (lactogenesis) induction in rabbits by reserpine, 113
Leukemia, induction of, and pulmonary tumors in mice by urethan, 29
Lipogenesis, difference in young and mature obese-hyperglycemic mice, 35, 36
  influence of compound A on, 38
  inhibition by drugs, 132
Lipolysis, impairment in obese-hyperglycemic mice, 35
α-Lipoproteinemia, 168
Liver disease, factor VII deficiency in, 151
LSD see Lysergic acid diethylamide
Lysergic acid diethylamide, ineffectiveness in altering steroid excretion in guinea pigs, 117

M

Malformations, in mice (129, BALB/c), due to 5-fluorouracil, 28
Man, drug metabolism as influenced by genetic factors in, 8
Mechanism of heredity, 15
Medmain (2-methyl-3-ethyl-5-dimethylaminoindole), antiserotonin action in dilute mice, 45
Menthol, influence on β-glucuronidase, 9
Mepazine, action on catecholamine storage, 124
Meprobamate, action on catecholamine storage, 124
  antiestrus effect in mice, 21
MER-29 see Triparanol

2-Mercaptoethylamine hydrochloride, influence on lifespan and spontaneous cancer in mice, 31
Mercuric chloride, retention in chickens, 129
2-C^{14}-Mevalolactone, incorporation into rat brain, 89
2-C^{14}-K-Mevalonate, incorporation into rat brain, 89
Metaglycodol (2-m-chlorophenyl-3-methyl-2,3-butanediol), antagonism of amphetamine in mice by, 23
2-Methoxyestradiol, influence on hairless mice, 51
3-Methylcholanthrene, tumor induction in mice by, 31
N-Methylformamide, influence on lifespan and spontaneous cancer in mice, 32
3'-Methyldimethylaminobenzene, effect on endoplasmic reticulum, 5
Mice, inbred strains of, 14
  hybrids, response to hormones, 24-27
Microbial systems, drug metabolism as influenced by genetic factors in, 8
Monase (d,1-α-ethyl-tryptamine), anticonvulsive action in mice suffering from audiogenic seizures, 46, 47
Monoamine oxidase, no change in, following adrenalectomy, in rats, 85
Morphine, excretion in dogs, 9
Mustards, nitrogen, in mice, 31
  sulfur, in mice, 31
Mutagenicity of drugs, 134, 135
Mutations, analogous, in man and animals, 2, 3
  definition of, 15
Myasthenia gravis, resemblance of thymuses from mice with autoimmune disease to, of man, 70
Myelation, cholesterol synthesis as parameter, in young rats, 88
  lack of and myelin degeneration, in dilute mice, 89

N

NADP(H_2) see Nicotinamide adenine nucleotide phosphate (reduced form)
Na-K ATPase, inhibition by digitalis alkaloid, 115

Subject Index

Na-Orinase [tolbutamide, 1-butyl-3-(*p*-tolylsulfonyl)urea], effects in normal and diabetic chinese hamsters, 101
NEFA *see* Non-esterified fatty acid(s)
Neurochemical strain differences, in mice, 32
Neuromuscular mutants, in mice, 55-62
Nicarbazin, as coccidiostat in chickens, 129
Nicotinamide adenine nucleotide phosphate dependence of steroid-metabolizing enzymes on, 88
  lack of reduction of acetoacetate in muscular dystrophy of mice by, 65
Nicotinic acid, lack of hypocholesterolemic effect in suckling rats, 90
Nicotine, selective influence on adrenal cholinesterase, 122
Nidroxyzone, antiestrus action in mice, 21
Nitrofurazone, as coccidiostat in chickens, 129
Nitrophenide, as chicken coccidiostat, 129
NPH *see* Insulin
Non-esterified fatty acid(s), elevation in hereditary obesity of rat, 92
Non-hemolytic jaundice, hereditary, in Gunn rats, 93, 94
Norepinephrine, brain contents of, in mice, 32
  influence in obese-hyperglycemic mice, 36
NSD 1034 [*N*-(*m*-hydroxybenzyl)-*N*-methylhydrazine], as inhibitor of decarboxylases, 47
Nucleotides and brain energy metabolism, 56

O

Obesity, hereditary, comparison between rat and mouse, 92, 93
  in mice (obese-hyperglycemic syndrome), 33-42
  in STR/IN mice, 78
Octamethyl pyrophosphoramide, sex-dependent toxicity in rats, 87
OMPA *see* Octamethyl pyrophosphoramide
Orinase *see* Na-Orinase
Osteoarthopathy, in STR/IN mice, 78
Osteogenesis imperfecta, in dogs, cats, and man, 120, 121

Osteoporosis, in obese-hyperglycemic mice, 40

P

Pentobarbital, sex-dependent action in rats, 87
  increase of enzyme activities for pentobarbital breakdown by, 87
Perphenazine, action on catecholamine storage, 124
  anti-estrus effect in mice, 21
Pharmacogenetics, definition of, 1
Pharmacologic responses, genetic dependence of, 8, 9
Phenaglycodol, increase of enzyme activities for phenobarbital breakdown by, 87
Phenformin (N^1-β-phenethylformamidinyliminourea hydrochloride), effects in normal and diabetic Chinese hamsters, 101
Phenobarbital, antagonism to amphetamine in mice, 23
Phenylalanine derivatives, inhibition of decarboxylases by, 45
Phenylalanine hydroxylase, inhibition in dilute mice, 45
Phenylketonuria, in mice, 42-48
Phenylmercuric acetate, retention in chickens, 129
Phosphatase *see* Glucose-6-phosphatase
Phospholipids, elevation in suckling rats, 90
Phosphorylase, adrenal stimulation by ACTH, 41
  hepatic, increase in obese-hyperglycemic mice, 39
Physiologic factors of drug metabolism (age, sex), 6, 7
Picrotoxin, sex-dependent toxicity in rats, 87
Piperacetazine [2-acetyl-10-(3-4-β-hydroxyethylpiperidino)phenothiazine], antagonism to amphetamine in mice, 23
Pituitary tumors in mice, 54, 55
Potassium arsenite, influence on life span and spontaneous cancer in mice, 31
Prediabetic lesions in Chinese hamsters, 104-109

Δ^5-Pregnenolone, influence on hairless mice, 51
Pregnenolone, influence on hairless mice, 51
Preputial gland tumor, of mice, pathway from lanosterol to cholesterol, 53
Primaquine, sensitivity to, due to lack of glucose-6-phosphatase, 167
Progestational activity, assay for, 132, 133
Progesterone-4-C^{14}, conversion to corticosterone and aldosterone, 40
 influence on hairless mice, 51
Promazine, action on catecholamine storage, in mice, 124
Prothrombin, bovine, purification, properties and composition, 156, 157
 complex deficiencies, 144
 rebound, post-parturient, 153
Protoporphyrin, defective synthesis in anemic mice, 67
PTC (Plasma Thromboplastin Component) see Hemophilia B
Purine, metabolism in Dalmatian dogs, 119
Pyrogallol, inhibition of catechol-O-methyl-transferase by, 37
Pyrophosphate, influence on lipolysis, 36
Pyrrolase, RNA-induced synthesis of, 166
Pyruvate, defective utilization in mouse muscular dystrophy, 64

Q

Quality, genetic control in mice, 12-18

R

Rabbits, breeds of, 109, 110
Rats, strains of, 80
Readily extractable cholesterol, as index of atherosclerosis susceptibility, 130
REC see Readily extractable cholesterol
Ribonucleic acid, biosynthesis of specific proteins by, 166
 glucose-6-phosphatase, 166
 pyrrolase, 166
 tryptophan, 166
RNA see Ribonucleic acid
Reserpiline, action on catecholamine storage, 124
Reserpine, induction of lactogenic response in rabbits by, 113
 sensitivity relative to accessory adrenal cortical tissue, 26
 strain differences and enhancement of toxicity of by adrenalectomy, 84, 85
Retention of copper, phenylmercuric acetate and mercury chloride in livers of chicken lines, 129
Retinal dystrophy, inheritance in mice, 75-77
Retinene, 11-cis, lack of utilization in retinal dystrophy of mice, 76

S

Salt resistance, in mice (IHB, NH), 19
Scopolamine, non-specific rabbit atropinesterase substrate, 112
 potential mutagenicity of,
Scurvy, in guinea pigs, 116
Serotonin (5HT), content in brain of mice, 32
 strain differences and enhancement of toxicity by adrenalectomy in rats, 84, 85
 toxicity of, in mice, 22
Serum proteins, in prediabetic Chinese hamsters, 107-109
 in spleenless mice, 74, 75
Sex, influence on drug metabolism, 67, 21, 23, 87
Similarities, physiologic, in mice, 21
SKF 525 A (diethylaminoethyl diphenylpropyl acetate), inhibition of pentobarbital-metabolizing enzymes by, 5, 87
Somatotropic hormones, of pituitary tumors, causing obese-hyperglycemic syndrome in parakeets, 101
Spastic mice, drug effects on, 62
Spleenlessness, hereditary, in mice, 74, 75
Steroidogenesis, in adrenal cortical tumors of mice, and interconversions of steroids in obese-hyperglycemic mice, 40
 in preputial gland tumors, 52, 53
Steroids, effects on body fat, 42
 excretion in guinea pigs, 116, 117
STH see Somatotropic hormone
Strychnine, sensitivity of mice to, 23
Stuart-Prower factor deficiency, 144-148

Subject Index

SU 3118 (syrosingopine) and SU 5171 [methyl-18-O-(3-N,N-dimethylaminobenzoyl) reserpate], strain differences in toxicity of, in adrenalectomized rats, 85
Succinoxidase, decrease in liver and adrenal cortex due to age, 6, 7
Sulfa drugs, intensification of neurological signs in hereditary jaundice of rats by, 93
Sulfaquinoxaline, as coccidiostat in chickens, 129
Suxamethonium sensitivity, due to lack of (pseudo-) cholinesterase, 167
Syrosingopine, SU 3118, release of catecholamines by, 85

T

Tangier disease, 168
Testosterone propionate, response of mice to, 24
Thalidomide, teratogenic action of, 136
  relation to glutamic acid metabolism, 136
Therapeutic approaches in atherosclerosis, 132
Therapy of clotting disorders by specific factor replacements, 154, 155
Thymus, resemblance in autoimmune disease of mice to human myasthenia gravis, 70
Thiourea, toxicity in rats, 83
Thyroid, activity in mice, 20
  initiation of function in various species, 131
  relation to atherosclerosis, 130
Thyroxine, injection causing accumulation of triiodothyronine, 55
Thyroxine and congeners, hypocholesterolemic effects in suckling rats, 90
Tolbutamide see Na-Orinase
Traits, inherited, of farm animals, 124-126
Transamidinase, glycine, elevation of renal, in mouse muscular dystrophy, 64
Trichlorobutanol see Chloretone
3,5,3-Triiodothyronine, accumulation in thyrotropic tumors after thyroxine injection, 55

Trimethandione, slight effect in spastic mice, 62
Triparanol (MER-29), ineffectiveness, in suckling rat hypercholesterolemia, 90
  influence on microsomes, 5
  as inhibitor of Δ^6-reductase, 53
Tryptophan, RNA-induced synthesis of, 166
Tryptophan metabolites, urinary, in mice, 45
Tumors, adrenal cortical, reduced enzymatic efficiency in, 54
  induction by drugs and metabolites causing, 29-31
  pituitary adrenotropic, control of secretions in, 54
  preputial gland, steroids production in, 52, 53
  thyrotropic, control of secretions in, 54
Tyrosinase, genetic influence on, 44

U

Urethan, induction of tumors in mice by, 29
Urinary calculi in mice, 74
Urinary excretion of amino acids in mice, 72-74
  of corticosteroids in guinea pigs, 116, 117
  of glucuronide in rats, 93, 94
  of morphine in dogs, 9
  of tryptophan metabolites in mice, 45
  of uric acid in Dalmatian dogs, 118, 119
Uterone (endometrial secretions), influence in fibrinolysis, 160, 161

V

Varieties and breeds of chickens, 128, 129
Vitamin K_1, relationship of, and analogs, on prothrombin complex, 151, 152

W

W-series, hereditary types of anemia in mice due to, 66-68

X

X-irradiation, influence on tumorigenesis in mice, 29

Z

Zymogram technique, 57
Zymosterol, 53